WMO/GEO Expert Meeting on an International Sand and Dust Storm Warning System 2007

AA002741

IOP Conference Series: Earth and Environmental Science
Volume 7

Barcelona, Spain
7 - 9 November 2007

ISBN: 978-1-5108-0204-9
ISSN: 1755-1307

Printed from e-media with permission by:

Curran Associates, Inc.
57 Morehouse Lane
Red Hook, NY 12571

Some format issues inherent in the e-media version may also appear in this print version.

Copyright© by the Institute of Physics
All rights reserved.

Printed by Curran Associates, Inc. (2015)

For permission requests, please contact the Institute of Physics
at the address below.

Institute of Physics
Dirac House, Temple Back
Bristol BS1 6BE UK

Phone: 44 1 17 929 7481
Fax: 44 1 17 920 0979

techtracking@iop.org

Additional copies of this publication are available from:

Curran Associates, Inc.
57 Morehouse Lane
Red Hook, NY 12571 USA
Phone: 845-758-0400
Fax: 845-758-2634
Email: curran@proceedings.com
Web: www.proceedings.com

Table of contents

Volume 7

WMO/GEO EXPERT MEETING ON AN INTERNATIONAL SAND AND DUST STORM WARNING SYSTEM
7–9 November 2007, Barcelona Supercomputing Center (BSC), Barcelona, Spain

Accepted papers received: 24 February 2009
Published online: 22 April 2009

PAPERS

The Global Distribution of Mineral Dust

I Tegen and K Schepanski

Leibniz Institute for Tropospheric Research, Leipzig, Germany

Email: itegen@mail.tropos.de

Abstract. Dust aerosol particles produced by wind erosion in arid and semi arid regions affect climate and air quality, but the magnitude of these effects is largely unquantified. The major dust source regions include the Sahara, the Arabian and Asian deserts; global annual dust emissions are currently estimated to range between 1000 and 3000 Mt/yr. Dust aerosol can be transported over long distances of thousands of kilometers, e.g. from source regions in the Saharan desert over the North Atlantic, or from the Asian deserts towards the Pacific Ocean. The atmospheric dust load varies considerably on different timescales. While dust aerosol distribution and dust effects are important on global scales, they strongly depend on dust emissions that are controlled on small spatial and temporal scales.

1. Regional distribution of dust aerosol

Soil dust aerosol impacts upon atmospheric radiation balance, oceanic and terrestrial biogeochemical processes, and atmospheric chemistry. High surface concentrations downwind of source areas adversely affect regional air quality. The magnitudes of those dust effects remain largely unquantified.

Dust aerosol particles are produced by wind erosion in arid and semi-arid continental regions. The main source regions include the Sahara desert and the Sahelian region in North Africa, the Arabian Peninsula, the Gobi and Taklamakan deserts in Asia, and the Australian and South American deserts [1]. The largest contribution to the global dust load is emitted from the North African (50-70%) and Asian deserts (10-25%). Dust emission is initiated when the surface wind friction velocity in a source region exceeds a threshold value, which depends on surface properties like surface roughness and soil moisture [2]. Dust emission fluxes depend on to the third power of the surface wind friction velocity and are particularly strong in areas that contain fine and loose sediments. A good agreement in maximal occurrence of dust over land surfaces observed by satellites and the location of enclosed topographic depressions has been observed [1].

When mobilized dust particles are vertically mixed into higher atmospheric levels (reaching up to several kilometers) they can be carried over distances of thousands of kilometers by strong wind systems. Such dust plumes can be clearly recognized over dark ocean surfaces by satellite remote sensing instruments. For example, large amounts of dust from the Sahara desert can be carried across the North Atlantic to North and Central America (e.g., [3]), or to Europe (e.g., [4]). Also, large amounts of Asian dust are carried over the North Pacific towards the mid-Pacific islands and North America (e.g. [5]). This long-range transport in dust usually takes place in well-defined elevated layers at several km height and cannot be characterized by surface measurements alone. Dust transport from Asia to Greenland (e.g., [6]) and from South America to Antarctica (e.g. [7]) has been found in ice cores and may provide information on past climate conditions.

© 2009 IOP Publishing Ltd

Dust particles are removed from the atmosphere either by precipitation scavenging, or by dry deposition. Close to dust source regions, the gravitational sedimentation is responsible most of the dry dust deposition, while wet deposition is responsible for a large part of the removal of far-traveled dust over remote ocean areas [8]. Atmospheric lifetimes of dust particles depend on their size and may range from few hours for particles larger than 20 μm that are quickly removed by gravitational settling, to some weeks for submicron particles that are mostly removed by wet deposition [9].

Estimates of the global dust budget can be obtained from extrapolation measurements, satellite remote sensing and global models of dust emission, transport and deposition. Model-based estimates range from 1000-3000 Mt/yr for global dust emissions and 8 to 35 Mt for the global annual average of the atmospheric dust load [10]. A major cause for the large uncertainty range in these estimates is caused by the lack of available observations near the source areas and the difficulties in quantitative retrievals of aerosol optical thickness above bright surfaces, which are characteristic for desert areas. The Total Ozone Mapping Spectrometer absorbing Aerosol Index (TOMS AI, [11]) has been a useful tool to detect the presence of dust aerosol above land surfaces, but quantitative retrievals of dust optical thicknesses are difficult to derive from this instrument. New retrieval methods like for example the 'Deep Blue' algorithm [12] that makes use of the low surface reflectivity in the blue part of the solar spectrum may improve quantitative dust retrievals over deserts in the future. The recent AEROCOM model intercomparison project that compared and evaluated about 15 global aerosol transport models [13] indicated that uncertainties in dust emission and initial dust size distribution were a major cause for discrepancies between the different models.

2. Temporal variability of dust aerosol

Dust aerosol strongly varies on different timescales. Interannual and decadal changes in dust emissions and dust loads are controlled by changes in relevant climate parameters like surface winds and the magnitude and distribution of precipitation, including modifications in vegetation cover resulting from climatic changes [14], and by modifications of surface properties. For example, the interannual variations in southern Saharan dust emissions are found to be correlated with the Sahelian rainfall deficit [3,15]. The disturbance of surface crusts as a consequence of human activities like cultivation in dry regions can lead to enhanced dust emissions (as observed for example during the 'dust bowl' events in the United States in the 1930s and 1950s). This contribution to the global dust load may be considerable in semi-arid regions, but the magnitude is very uncertain on the global scale. While [16] estimates a contribution of this enhanced dust of up to 50% to the global dust load, recent investigations conclude that this contribution should be much smaller, probably less than 20% globally [17].

Seasonal changes in the atmospheric dust load are well characterized; these follow the seasonal changes in meteorology and vegetation phenology (figure 1). A well-known seasonal dust feature is for example the maximum in Asian dust emissions in the Gobi and Taklamakan deserts and eastward transport toward the Pacific following passages of cold fronts in spring ('Kosa' events) that cause strong winds in dust source regions. Another dominant feature is the seasonal shift of the Saharan dust plume that crosses the North Atlantic. In the northern hemisphere winter, Saharan dust is transported south-westward and can reach as far as the Amazon basin, while in summer, Saharan dust is transported in westerly direction as far as the United States. High dust concentrations are simulated over the Arabian peninsula in late spring (during the season of 'Shamal' winds [18]), while the maximum dust concentrations over the Arabian Sea occurs later during the summer months. The location of a dust maximum around the Caspian and Aral seas in spring agrees with the occurrence of dust storms in this region of Central Asia. In the southern hemisphere, dust is emitted and transported from desert regions in South America and Australia. The dust production and transport from these regions is generally smaller than from northern hemisphere sources.

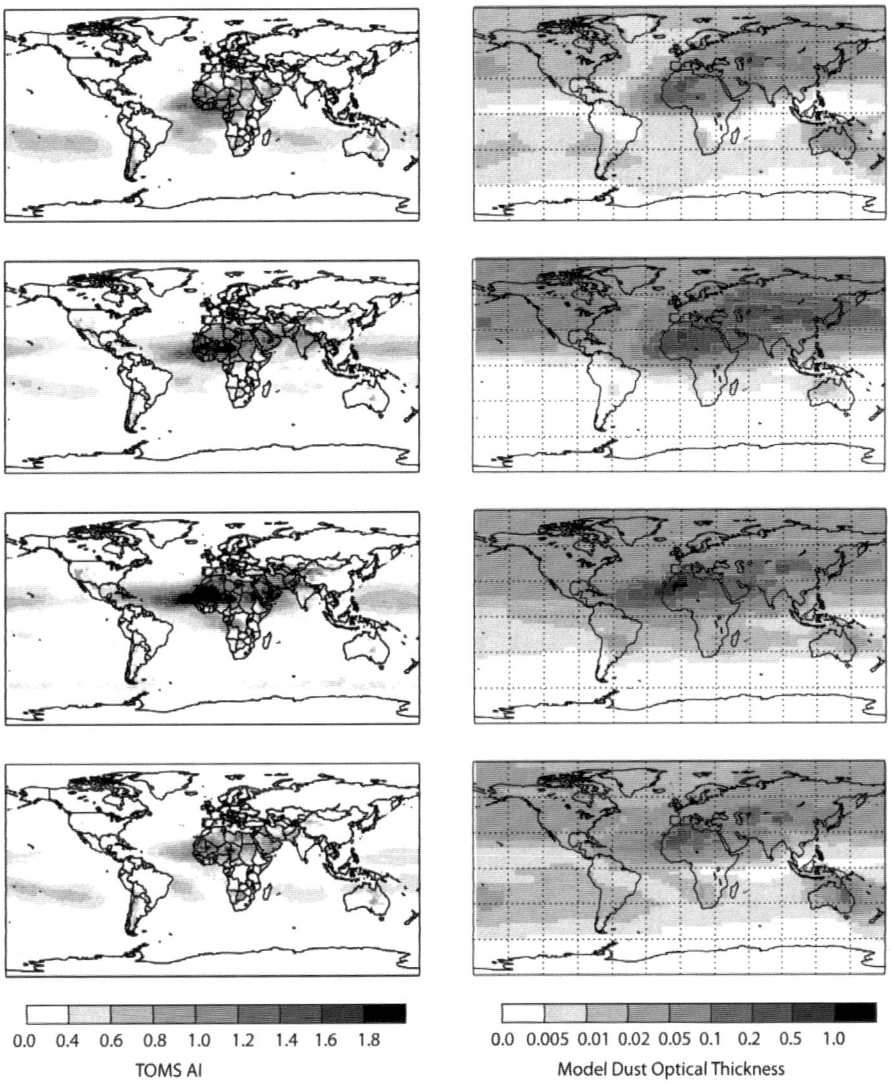

Figure 1. Seasonal changes in global dust distribution represented by 10-year average of seasonal TOMS AI (left) and modeled dust optical thickness (right) from a global offline transport model [19].

While dust transport is a global-scale phenomenon, it is controlled by small-scale regional processes. In regions downwind of large dust source regions, forecasting dust events that impact on air quality is important. Daily and hourly variability in dust emission and transport are relevant for such forecast studies. Dust emission events are often related to convective systems or the passage of cold fronts. Convective events causing dust emissions include dry convective events (dust devils) that are mainly of local importance [20]. Thunderstorms forming in unstable conditions may cause vigorous surface winds. Sometimes they are accompanied by intense horizontal vortices with violent surface winds, so-called squall lines. In the arid regions of southwestern US and in particular the southern Sahara such squall lines are frequently observed in the summer months, and are a major cause for dust production [21]. In many regions in the Sahara, e.g. in the Bodélé depression dust emission is caused by turbulent downward mixing of momentum from a low-level jet, a process described by [22].

3. Saharan dust sources

To characterize dust emissions at sub-daily timescales, imagery from geostationary satellites plays an important role. Recently, 15-minute images of the Meteosat Second Generation (MSG) infra-red (IR) difference dust index images were used to localize dust source activation in the Sahara desert for a full year [23]. The MSG dust index is composed of the differences of the brightness temperatures at the wavelengths 8.7 μm, 10.8 μm and 12.0 μm. With a sampling rate of 15 minutes and a spatial resolution of 3 km at nadir, this index indicates the presence and movement of dust over the Saharan desert surface at very high spatiotemporal resolution and is therefore useful for identifying the origin of individual dust events. The thus derived main Saharan dust source areas not always agree with source areas derived from daily TOMS AI retrievals [1]. Topographic depressions that were interpreted as strong dust source areas in the earlier analysis of daily dust indices are only partly identified to be important dust sources in the new analysis at higher temporal resolution. Instead, high frequencies of dust emission are found at foothills of mountain areas where fine sediment is available in wadis and fluvial outwash, as in the foothills of the Massif l'Air and the foothills of the Red Sea Mountains [2]. Many models of dust emission, transport and deposition at global [19,24,25] and regional scales [26] prescribe topographic depressions as preferential sources for dust emission, which may lead to model errors, as shown for the example of the results of dust source activation frequency for the Saharan desert computed with the model of [19] compared to the activation frequencies by [23] (figure 2). Nevertheless, in spite of the mismatch in the precise determination of dust emission areas, model studies of individual dust events have been successful in simulating the spatio-temporal dust distribution of dust transported far downwind, e.g. in cases of dust transport from the Sahara desert towards Europe (e.g. [26,27]). While a precise characterization of dust source areas and emission fluxes is crucial for studies of feedbacks of dust and climate, the importance for regional forecast studies remains to be clarified.

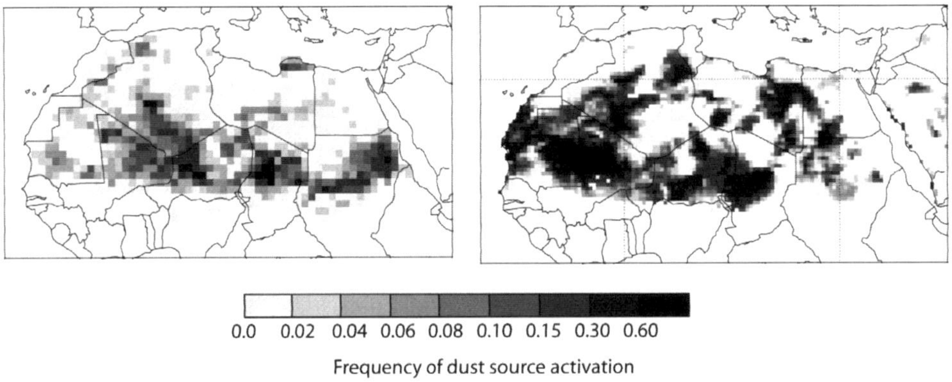

Figure 2. Annual Saharan dust source activation based on Meteosat-8/SEVIRI 15-min IR dust index retrievals [23] (left) and model results of dust source activation from model results by [19] (right).

References

[1] Prospero, J. M., P. Ginoux, O. Torres, S. E. Nicholson, and T. E. Gill 2002 Environmental characterization of global sources of atmospheric soil dust identified with the Nimbus 7 Total Ozone Mapping Spectrometer (TOMS) absorbing aerosol product, *Rev. Geophys.*, **40**(1), 1002.

[2] Marticorena, B., G. Bergametti, B. Aumont, Y. Callot, C. N'Doumé, and M. Legrand 1997 Modeling the atmospheric dust cycle 2. Simulation of Saharan dust sources, *J. Geophys. Res.*, **102**(D4), 4387-4404.

[3] Prospero, J. M. and Lamb, P. J. 2003 African droughts and dust transport to the Caribbean: Climate change implications. *Science* **302**, 1024-1027.

[4] Barkan, J., P. Alpert, H. Kutiel, and P. Kishcha 2005 Synoptics of dust transportation days from Africa toward Italy and central Europe, *J. Geophys. Res.,* **110**, D07208, doi:10.1029/2004JD005222.

[5] Sassen, K., P. J. DeMott, J. M. Prospero, and M. R. Poellot 2003 Saharan dust storms and indirect aerosol effects on clouds: CRYSTAL-FACE results, *Geophys. Res. Lett.,* **30**(12), doi:10.1029/2003GL017371.

[6] Biscaye, P. E., Grousset, F. E., Revel, M., VanderGaast, S., Zielinski, G. A., Vaars, A., Kukla, G. 1997 Asian provenance of glacial dust (stage 2) in the Greenland Ice Sheet Project 2 Ice Core, Summit, Greenland. *J. Geophys. Res.* **102**, 26765-26781.

[7] Basile, I., Grousset, F. E., Revel, M., Petit, J. R., Biscaye, P. E., Barkov, N. I. 1997 Patagonian origin of glacial dust deposited in East Antarctica (Vostok and Dome C) during glacial stages 2, 4 and 6. *Earth Planet. Sci. Lett.* **146**, 573-589.

[8] Jickells, T. D, An, Z. S., Andersen, K. K., Baker, A. R., Bergametti, G., Brooks, N., Cao, J. J., Boyd, P. W., Duce, R. A., Hunter, K. A., Kawahata, H., Kubilay, N., LaRoche, J., Liss, P. S., Mahowald, N. M., Prospero, J. M., Ridgwell, A. J., Tegen, I., Torres, R., 2005 Global iron connections between desert dust, ocean biogeochemistry, and climate. *Science* **308**, 67-71.

[9] Seinfeld, J. H. and Pandis, S. N. 1997 Atmospheric Chemistry and Physics: From Air Pollution to Climate Change, John Wiley, New York.

[10] Zender, C. S., R. L. Miller, and I. Tegen 2004 Quantifying mineral dust mass budgets:Terminology, constraints, and current estimates, *EOS Transactions*, **85**(48), 509-512.

[11] Herman, J. R., Bhartia, P. K., Torres, O., Hsu, C., Seftor, C., Celarier, E. 1997 Global distribution of UV-absorbing aerosols from Nimbus 7/TOMS data. *J. Geophys. Res.* **102**, 16,911-16,922.

[12] Hsu, N. C., S. C. Tsay, M. D. King, and J. R. Herman, 2006 Deep blue retrievals of Asian aerosol properties during ACE-Asia. IEEE Trans. *Geosci. Remote Sens.,* **44**, 3180-3195.

[13] Textor, C., et al. (2006), Analysis and quantification of the diversities of aerosol life cycles within AeroCom, *Atmos. Chem. Phys.,* **6**, 1777-1813, 2006.

[14] Mahowald, N. M. 2007 Anthropocene changes in desert area: Sensitivity to climate model predictions, *Geophys. Res. Lett.,* **34**, L18817, doi:10.1029/2007GL030472.

[15] Chiapello, I., C. Moulin, and J. M. Prospero 2005 Understanding the long-term variability of African dust transport across the Atlantic as recorded in both Barbados surface concentrations and large-scale Total Ozone Mapping Spectrometer (TOMS) optical thickness, *J. Geophys. Res.,* **110**(D18).

[16] Intergovernmental Panel on Climate Change (IPCC) 2001.(2001) Climate Change 2001: The Scientific Basis (ed. J. T. Houghton, Y. Ding, D. J. Griggs, M. Noguer, P. J. van der Linden, X. Dai, K. Maskell and C. A. Johnon), Cambridge Univ. Press, New York.

[17] Intergovernmental Panel on Climate Change (IPCC) 2007. (2007), Climate Change 2007: The Physical Science Basis, (ed. Solomon, S., D. Qin, M. Manning, M. Marquis, K. Averyt, M. M. B. Tignor, and H. L. Miller) 989, Cambridge University Press, U.K.

[18] Pye, K. 1987 Aeolian dust and dust deposition, Academic Press Inc., London.

[19] Tegen, I., Werner, M., Harrison, S. P., Kohfeld, K. E. 2004 Relative importance of climate and land use in determining present and future global soil dust emission. *Geophys. Res. Lett.* **31**, L05105, doi:10.1029/2003GL019216.

[20] Balme, M., S. Metzger, M. Towner, T. Ringrose, R. Greeley, and J. Iversen 2003 Friction wind speeds in dust devils: A field study, *Geophysical Research Letters,* **30**(16).

[21] Shao, Y. P. 2000 Physics and Modelling of Wind Erosion, Kluwer Academic Publishers, Dordrecht, The Netherlands.

[22] Washington, R., M. C. Todd, S. Engelstaedter, S. Mbainayel, and F. Mitchell 2006 Dust and the low-level circulation over the Bodélé Depression, Chad: Observations from BoDEx 2005, *J. Geophys. Res.,* **111**(D3).

[23] Schepanski, K., I. Tegen, B. Laurent, B. Heinold, and I. Macke 2007 A new Saharan dust source activation frequency map derived from MSG-SEVIRI IR-channels, *Geophys. Res. Lett.,* **34**, L18803.

[24] Ginoux, P., M. Chin, I. Tegen, J. Prospero, B. Holben, O. Dubovik, and S.-J. Lin, 2001 Global simulation of dust in the troposphere: Model description and assessment *J. Geophys. Res.,* **106**, 20,255– 20,274.

[25] Zender, C. S., D. Newman, and O. Torres 2003 Spatial heterogeneity in aeolian erodibility: Uniform, topographic, geomorphic, and hydrologic hypotheses, *J. Geophys. Res.,* **108**(D17).

[26] Heinold, B., J. Helmert, O. Hellmuth, R. Wolke, A. Ansmann, B. Marticorena, B. Laurent, and I. Tegen 2007 Regional modeling of Saharan dust events using LM-MUSCAT: Model description and case studies, *J. Geophys. Res.,* **112**(D11). D11204.

[27] Pérez, C., S. Nickovic, G. Pejanovic, J. M. Baldasano, and E. Ozsoy 2006 Interactive dust-radiation modeling: A step to improve weather forecasts, *J. Geophys. Res.*, **111**(D16).

IOP Conf. Series: Earth and Environmental Science 7 (2009) 012002 doi:10.1088/1755-1307/7/1/012002

EARLINET coordinated lidar observations of Saharan dust events on continental scale

G Pappalardo[1], A Papayannis[2], J Bösenberg[3], A Ansmann[4], A Apituley[5], L Alados Arboledas[6], D Balis[7], C Böckmann[8], A Chaikovsky[9], A Comeron[10], O Gustafsson[11], G Hansen[12], V Mitev[13], L Mona[1], D Nicolae[14], M Rita Perrone[15], A Pietruczuk[16], M Pujadas[17], J-P Putaud[18], F Ravetta[19], V Rizi[20], V Simeonov[21], N Spinelli[22], D Stoyanov[23], T Trickl[24] and M Wiegner[25]

[1]Consiglio Nazionale delle Ricerche -Istituto di Metodologie per l'Analisi Ambientale CNR-IMAA, Potenza, Italy
[2]National Technical University of Athens, Department of Physics, Athens, Greece
[3]Max-Planck-Institut für Meteorologie, Hamburg, Germany
[4]Institut für Troposphärenforschung, Leipzig, Germany
[5]Rijksinstituut voor Volksgezondheid en Milieu, Bilthoven, The Netherlands
[6]Universidad de Granada, Granada, Spain
[7]Aristoteleio Panepistimio, Thessalonikis, Greece
[8]Institut für Mathematik der Universität Potsdam, Germany
[9]Institute of Physics, National Academy of Sciences, Minsk, Bjelarus
[10]Universitat Politècnica de Catalunya, Barcelona, Spain
[11]Swedish Defence Research Agency (FOI), Linköping, Sweden
[12]Norwegian Institute for Air Research at the Polar Environmental Centre, Tromsö, Norway
[13]CSEM, Centre Suisse d'Electronique et de Microtechnique SA, Neuchâtel, Switzerland
[14]National Institute of R&D for Optoelectronics, Magurele-Bucharest, Romania
[15]Università degli Studi del Salento, Department of Physics, Lecce, Italy
[16]Institute of Geophysics, Polish Academy of Sciences, Warsaw, Poland
[17]Centro de Investigaciones Energéticas, Medioambientales y Tecnológicas, Madrid, Spain
[18]EC Joint Research Centre, Ispra (VA), Italy
[19]Université Pierre et Maris Curie-Institut Pierre Simon Laplace, Paris, France
[20]Università degli Studi dell'Aquila - Dipartimento di Fisica - CETEMPS, L'Aquila, Italy
[21]Ecole Polytechnique Fédérale de Lausanne, Switzerland
[22]Consorzio Nazionale Interuniversitario per le Scienze Fisiche della Materia, Napoli, Italy
[23]Institute of Electronics, Bulgarian Academy of Sciences, Sofia, Bulgaria
[24]Forschungszentrum Karlsruhe IMK-IFU, Garmisch-Partenkirchen, Germany
[25]Ludwig-Maximilians-Universität, München, Germany

Email: pappalardo@imaa.cnr.it

© 2009 IOP Publishing Ltd

IOP Conf. Series: Earth and Environmental Science **7** (2009) 012002 doi:10.1088/1755-1307/7/1/012002

Abstract. EARLINET, the European Aerosol Research Lidar Network, is the best tool to investigate the horizontal and vertical transport of aerosols over Europe. Within the network, particular attention is devoted to Saharan dust events monitoring. An alert system has been established in order to perform devoted measurements in case of intrusions of desert particles on European continent. Starting from data collected within EARLINET since May 2000, a first statistical analysis of the aerosol vertical distribution on European scale during Saharan dust outbreaks, has been performed. These results highlights the fundamental role that EARLINET can have for the study of impact of Saharan dust on European scale. The current 5-year EU project EARLINET-ASOS, started in March 2006, will enhance the operation of the network through the improvement of the instruments and of the temporal coverage, and of the data analysis procedures.

1. Introduction

According to the latest report of the Intergovernmental Panel on Climate Change (IPCC) [1], there is still a large uncertainty about the estimation of the effect on climate forcing of anthropogenic and natural dust. A main cause of these actual large uncertainties is the lack of an extensive climatology based on dust observations [2]. In this context, EARLINET (European Aerosol Research LIdar NETwork) [3] can strongly contribute to quantify dust vertical and horizontal distribution by means of vertical profiles of desert dust optical properties. Before EARLINET, Saharan dust intrusions over Mediterranean area were systematically studied only by passive satellite-borne sensors [4,5,6,7] and ground-based sunphotometers [8], without any information about the vertical distribution. The few papers present in literature about the desert dust vertical distribution over the Mediterranean Basin, were typically based on not-systematic measurements [e.g. 9,10,11].

Since May 2000, EARLINET provides systematic observations of Saharan dust events over Europe. The main result up to know is the establishment of a significant database of vertical profiles of Saharan dust optical properties over Europe.

2. EARLINET

EARLINET is a coordinated network of stations that make use of advanced lidar methods for the vertical profiling of aerosols. At present, the network includes 25 lidar stations distributed over Europe, as shown in figure 1.

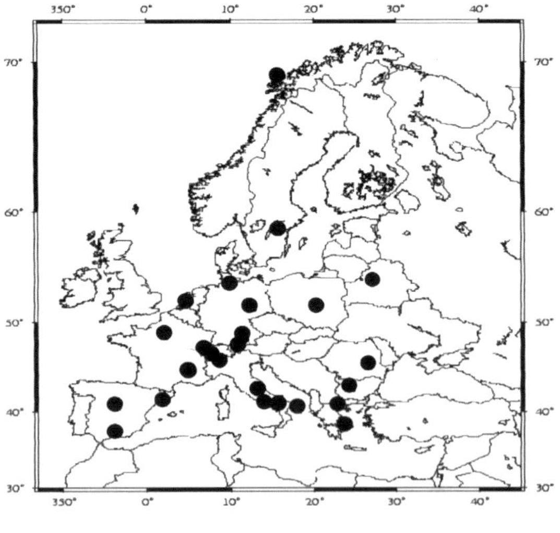

List of EARLINET stations

BELARUS:	Minsk (mi)
BULGARIA:	Sofia (sf)
FRANCE:	Palaiseau (pl), Observatoire de Haute Provence (hp)
GERMANY:	Hamburg (hh), Leipzig (lc), Munich (mu), Garmisch-Partenkirchen (gp)
GREECE:	Thessaloniki (th), Athens (at)
ITALY:	Ispra (is), L'Aquila (la), Napoli (na), Potenza (po), Lecce (lc)
NORWAY:	Andøya (an)
POLAND:	Belsk (be)
ROMANIA:	Bucharest (bu)
SWITZERLAND:	Lausanne (ls), Neuchâtel (ne)
SPAIN:	Granada (gr), Madrid (ma), Barcelona (ba)
SWEDEN:	Linkoping (lk)
THE NETHERLANDS:	Cabauw (ca)

Figure 1. Map of Europe with all the EARLINET lidar stations.

EARLINET lidar observations are performed at each station on a regular schedule of one daytime measurement per week around noon, when the boundary layer is usually well developed, and two night time measurements per week, in low background light conditions, in order to perform Raman extinction measurements. Further observations are addressed to monitor special events such as Saharan dust outbreaks, forest fires, photochemical smog and volcano eruptions [3].

Special care has been taken to assure data of highest possible quality. Therefore all network stations participated in intercomparisons both at instrument and algorithm levels with standardised procedures [12,13,14].

EARLINET was established in 2000 with the main goal to provide a comprehensive, quantitative, and statistically significant data base for the aerosol distribution on a continental scale. On March 2006, the 5-year EC Project EARLINET-ASOS (Advanced Sustainable Observation System) started on the base of the EARLINET infrastructure. This project will contribute to the improvement of continuing observations and methodological developments that are urgently needed to provide the multi-year continental scale data set necessary to assess the impact of aerosols on the European and global environment and to support current and future satellite missions. The main objective is the improvement of the EARLINET infrastructure, resulting in a better spatial and temporal coverage of the observations, continuous quality control for the complete observation system, and fast availability of standardised data products.

3. EARLINET special observations: Saharan dust events

During Saharan dust outbreaks, the lofted particles reach very often the Europe continent due to the proximity of the source region and to the prevailing meteorological conditions. Dust particles can be transported at different altitude ranges affecting the local aerosol content when it reaches very low altitude and the planetary boundary layer, while, if lofted in the free troposphere, dust particles can be transported over long distances, penetrating deeply into Northern Europe.

Thanks to its capability to provide high quality dust vertical profiles and to its geographical coverage, EARLINET is an optimal observation tool for studying Saharan dust events and their impact over Europe. With its current 25 stations, EARLINET covers almost the whole European continent from Madrid to Belsk and from Andøya to Athens. In particular, 12 stations are actually located in Southern Europe (namely below 45° N) from Spain to Romania through Italy and Greece, providing a full and detailed coverage of the passing area of dust from source region toward the whole Europe. Within EARLINET, 7 stations are multi-wavelength Raman lidars capable to provide extinction and backscatter profiles at different wavelengths. This allows to obtain information about aerosol microphysical properties and, in conjunction to the EARLINET capability to perform measurements on the whole continent, to study the aerosol modification processes occurring during the transport over Europe.

A suitable observing methodology has been established within the network, based on Saharan dust forecasts distributed to all EARLINET stations by the NTUA (National Technical University of Athens) group, which coordinates the monitoring of the dust outbreaks within the network. The dust forecast is based on the operational outputs (aerosol dust load) of the DREAM (operated in the past in Malta at the Centre on Insular Coastal Dynamics (ICOD) and now at the Barcelona Supercomputing Center, Barcelona, Spain: http://www.bsc.es/projects/earthscience/DREAM/) and the Skiron models (operated at Athens University, Athens, Greece: http://forecast.uoa.gr) [15,16]. The alerts are diffused 24 to 36 hours prior to the arrival of dust aerosols over the EARLINET sites, so that each participating station can arrange the special measurements. Typically long runs of measurements are performed at the EARLINET stations in order to investigate also the temporal evolution of the dust events. To identify the origin of the observed aerosol layer, the 4-day backward trajectories provided by the German Weather Service (DWD) are used. The DWD provides these back-trajectories at each EARLINET lidar station for two arrival times per day and for six arrival pressure levels between 200 and 975 hPa [3]. Additionally HYSPLIT back-trajectory analysis provided by the NOAA Air Resources Laboratory (ARL) are used too. Besides air mass back-trajectory analysis, satellite data analysis, like TOMS Aerosol Index and MODIS optical depth, have been utilized as ancillary information for identifying the cases in which air masses coming from the Sahara region carry on lofted desert dust particles towards EARLINET sites.

EARLINET data have already conducted to studies of Saharan dust transport over Europe [17,18] and to first climatological studies on Saharan dust events based on systematic observations [19, 20,21, 22,23].

A first study on the impact of Saharan dust on European aerosol content has been performed on the base of all EARLINET data acquired during the first period of operation of the network (2000-2002) [24]. During this period about 150 days of Saharan dust outbreaks over Europe were monitored. In most of the cases, episodes lasted from 1 to 5 days, while few of them lasted even up to 7-10 days. Multiple aerosol layers of variable thickness, ranging typically from 1500 m to 3400 m, were observed at altitudes between 1100 and 9000 m a.s.l. . Traces of dust particles have been observed in some cases up to 10000 m in Northern Europe.

The largest number of cases has been recorded from late spring until early autumn months, while only few cases were observed during the winter period. However, the dust observations during autumn and winter cannot be negligible at least for Southern (S) and South-Eastern (SE) European regions. Mean number of dust episodes observed per season is much higher in the S and SE regions as shown in Figure 2, mainly because of the proximity of these regions to the source region and to prevailing wind direction. The Central (C) and South-Western (SW) sectors still exhibit a quite significant number of events, while the remaining sectors, North (N), North-Eastern (NE) and Western (W), show low dust transport activity.

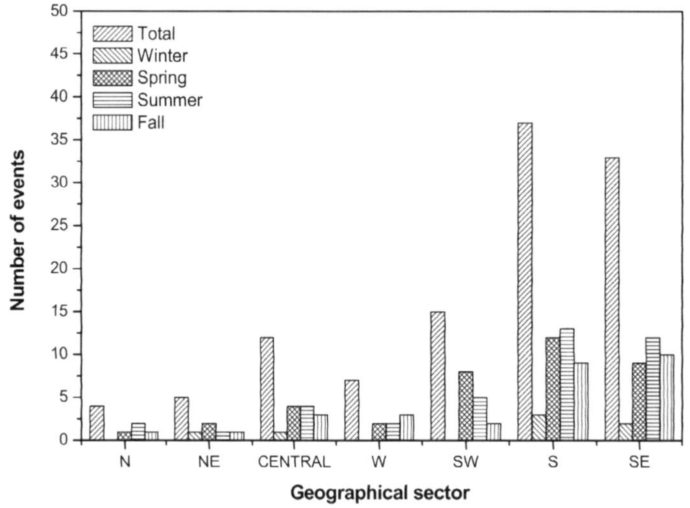

Figure 2. Seasonal variability of Saharan dust observed by EARLINET stations during May 2000-December 2002 period per sector over the European continent (north (hh, kb, le, lk) , north-eastern (be, mi), central (gp, ju, ne, mu), western (pl), south-western (ba), south (la, lc, na, po) and south-eastern (at, sf, th) Europe).

The mean altitude of the core of the dust layer is typically observed at about 3500 m above sea level (a.s.l.). Exceptions are found for the Barcelona and Lecce stations, where the dust layer is located around 2500 m, and for Minsk where it rises up to 6000 m. Over southern and south-eastern Europe a high variability is observed for the dust layer thickness, ranging between 700 m for Thessaloniki up to 3340 m for Potenza, without significant latitude dependence.

Regarding optical properties, mean AOD (Aerosol Optical Depth) values inside the dust lofted layers ranged over Europe from 0.1 to 0.25 (at 355 or 351 nm). A large variability in the AOD is observed for all the stations related to the large variability of dust origin and transport processes.

4. Future perspectives

EARLINET is continuing its activity and further lidar stations are going to join the network. In particular, the EARLINET database of Saharan dust vertical profiles is continuously extending.

In the next years, EARLINET-ASOS will enhance the operation of the network concerning also the dust observations. Instruments will be improved through more automated and unattended lidar systems, allowing a better temporal coverage of the observations. Moreover, the optimisation of data processing will provide an important step for near real time data delivery.

EARLINET data will be used for dust model validation: a systematic quantitative comparison with the DREAM model is already planned using lidar data from the whole network.

The future availability of near real time vertical profiles will be an important step for the assimilation of lidar data in dust models and this will result in a strong contribution to the implementation of the World Weather Research Programme and the Global Atmospheric Watch (WWRP/GAW) Sand and Dust Storm Warning System (SDS WS).

On global scale, EARLINET will give a strong contribution to the implementation of the GAW Atmospheric LIdar Observation Network (GALION) [25].

Acknowledgments

The EARLINET-ASOS project is funded by the EC under grant RICA-025991. We acknowledge the German Weather Service for the air mass back-trajectories and the NOAA Air Resources Laboratory (ARL) for the provision of the HYSPLIT back-trajectories. The NASA/GSFC is gratefully acknowledged for the EP/TOMS Aerosol Index and MODIS images. The authors also thank the Maltese Euromediterranean Center (ICoD), and the Barcelona Supercomputing Center for Dust Regional Atmospheric Model (DREAM) for the provision of dust forecast, and the atmospheric modelling and weather forecasting group of the University of Athens for the SKYRON dust forecast provision.

References

[1] Forster, P., et al. 2007 Changes in Atmospheric Constituents and in Radiative Forcing. In: Climate Change 2007: The Physical Science Basis. Contribution of Working Group I to the Fourth Assessment Report of the Intergovernmental Panel on Climate Change [Solomon, S., D. Qin, M. Manning, Z. Chen, M. Marquis, K.B. Averyt, M.Tignor and H.L. Miller (eds.)]. Cambridge University Press, Cambridge, United Kingdom and New York, NY, USA.

[2] Sokolik, I.N., et al. 2001 Introduction to special section: outstanding problems in quantifying the radiative impacts of mineral dust, *J. Geophys. Res.*, **106**, 18015-18027.

[3] Bösenberg, J, et al. 2003 EARLINET: A European Aerosol Research Lidar Network to Establish an Aerosol Climatology, MPI-Report No. 348, Hamburg, Germany.

[4] Antoine, D., and D. Nobileau 2006 Recent increase of Saharan dust transport over the Mediterranean Sea, as revealed from ocean color satellite (SeaWiFS) observations, *J. Geophys. Res.*, **111**, D12214, doi:10.1029/2005JD006795.

[5] F. Barnaba and G. P. Gobbi 2004 Aerosol seasonal variability over the Mediterranean region and relative impact of maritime, continental and Saharan dust particles over the basin from MODIS data in the year 2001, *Atmos. Chem. Phys.*, **4**, 2367–2391.

[6] Moulin C., et al. 1998, Satellite climatology of African dust transport in the Mediterranean atmosphere, *J. Geophys. Res.*, **103**, D11, 13137-13144.

[7] Dulac F., et al. 1992, Assessment of the African airborne dust mass over the western Mediterranean Sea using Meteosat data, *J. Geophys. Res.*, **97**(D2), 2489-2506.

[8] Holben, B. N., et al. 2001 An emerging ground-based aerosol climatology: Aerosol optical depth from AERONET, *J. Geophys. Res.*, **106**, 12,067–12,098.

[9] di Sarra A., et al. 2002, Effects of desert dust and ozone on the ultraviolet irradiance at the Mediterranean island of Lampedusa during PAUR II, *J. Geophys. Res.*, **107**(D18), 8135, doi:10.1029/2000JD000139.

[10] Pace, G., et al. 2005, Aerosol optical properties at Lampedusa (Central Mediterranean) – 1. Influence of transport and identification of different aerosol types, *Atmos. Chem. Phys. Discuss.*, **5**, 1-41.

[11] Gobbi, G.P., et al. 2004 The vertical distribution of aerosols, Saharan dust and cirrus clouds in Rome (Italy) in the year 2001, *Atmos. Environ.*, **41**, 261-275.

[12] Böckmann, C. et al. 2004 Aerosol lidar intercomparison in the framework of the EARLINET project. 2. Aerosol backscatter algorithms, *Appl. Opt.*, **43**, 977-989.

[13] Matthias, V. et al. 2004) Aerosol lidar intercomparison in the framework of the EARLINET project. 1. Instruments, *Appl. Opt.*, **43**, 961-976.

[14] Pappalardo, G., et al. 2004 Aerosol lidar intercomparison in the framework of the EARLINET project. 3. Raman lidar algorithm for aerosol extinction, backscatter and lidar ratio, *Appl. Opt.* **43**, 5370-5385.

[15] Kallos, G. and SKIRON group , 1998: The SKIRON forecasting system: VOL. I: Preprocessing ISBN 960-8468-15-9; VOL. II: Model description ISBN 960-8468-16-7; VOL. III: Numerical techniques ISBN 960-8468-17-5; VOL. IV: Paralellization ISBN 960-8468-18-3; VOL. V: Postprocessing ISBN 960-8468-19-1; VOL. VI: Procedures 960-8468-20-5.

[16] Nickovic et al., 2001 A model for prediction of desert dust cycle in the atmosphere. *Journal of Geophysical Research,* **106**, D16, 18113-18129.

[17] Ansmann A., et al. (2003), Long-range transport of Saharan dust to northern Europe: The 11-16 October 2001 outbreak observed with EARLINET, *J. Geophys. Res.*, **108** (D24), 4783, doi:10.1029/2003/2003JD003757.

[18] Pérez, C., et al. 2006 A long Saharan dust event over the western Mediterranean: Lidar, Sun photometer observations, and regional dust modeling, *J. Geophys. Res.*, **111**, D15214, doi:10.1029/2005JD006579.

[19] Mattis, I., et al. 2002 Dual-wavelength Raman lidar observations of the extinction-to-backscatter ratio of Saharan dust, *Geophys. Res. Lett.*, **29**, doi:10.1029/2002GL014721.

[20] De Tomasi, F., A. Blanco, and M.R. Perrone 2003 Raman lidar monitoring of extinction and backscattering of African dust layers and dust characterization, *Appl. Opt.*, **42**, 1699–1709.

[21] Balis, D., et al. 2004 Optical properties of Saharan dust layers as detected by a Raman lidar at Thessaloniki, Greece, *Geophys. Res. Lett.*, **31**, L13104, doi:10.1029/2004GL019881.

[22] Papayannis, A., et al. 2005 Measurements of Saharan dust aerosols over the Eastern Mediterranean using elastic backscatter-Raman lidar, spectrophotometric and satellite observations in the frame of the EARLINET project, *Atmos. Chem. Phys.*, **5**, 2065-2079.

[23] Mona, L., et al. 2006 Saharan dust intrusions in the Mediterranean area: Three years of Raman lidar measurements, *J. Geophys. Res.*, **111**, D16203, doi:10.1029/2005JD006569.

[24] Papayannis, A., et al. 2008 Systematic lidar observations of Saharan dust over Europe in the frame of EARLINET (2000-2002), *J. Geophys. Res.* (in press).

[25] Bösenberg, J. and R. Hoff (2008). GAW Aerosol Lidar Observation Network (GALION), WMO GAW Report (in press), WMO, Geneva, Switzerland.

IOP Conf. Series: Earth and Environmental Science **7** (2009) 012003 doi:10.1088/1755-1307/7/1/012003

Observation of Asian dust and air-pollution aerosols using a network of ground-based lidars (ADNet): Realtime data processing for validation/assimilation of chemical transport models

N Sugimoto[1] and I Uno[2]

[1] Atmospheric Environment Division, National Institute for Environmental Studies (NIES), Tsukuba, Japan
[2] Research Institute for Applied Mechanics, Kyushu University, Fukuoka, Japan

E-mail: nsugimot@nies.go.jp

Abstract. Network observations of Asian dust and air-pollution aerosols are being conducted with automatic two-wavelength (1064nm, 532nm) polarization (532nm) lidars. Currently, the lidars are operated at 20 locations including cooperative stations. Aiming at real-time validation/assimilation of chemical transport models, an automatic data processing system was developed. A method using the lidar depolarization ratio is employed to estimate the extinction coefficient profiles of Asian dust and spherical aerosols. At the same time, the attenuated backscattering coefficients at 1064nm and 532nm, and the total depolarization ratio at 532nm are derived. A data assimilation system based on 4-dimensional variational (4DVAR) method was developed for the lidar-network Asian dust data. In the assimilation, the dust emission factor was used as the control parameter. The results show the data assimilation is effective not only for improving the model results but also for estimating the emission in the dust source region.

1. Lidar network for Asian dust monitoring

We conduct the lidar network observations in cooperation with various research organizations and universities under several research programs [1,2,3]. Table 1 lists the locations of the lidar observation sites. A part of the network is operated as the Asian-dust monitoring network of Ministry of the Environment, Japan, which is related to the program of Asian Development Bank (ADB) and Global Environment Facility (GEF) [4]. Also, a part of the lidar network is operated in the radiation observation network, SKYNET [http://atmos.cr.chiba-u.ac.jp/].

The lidars used in the network are two-wavelength (532 nm, 1064 nm) Mie-scattering lidars with the depolarization ratio measurement function at 532 nm. They are operated automatically, and the 5-minute averaged lidar profiles are recorded every 15 minutes in the continuous observations. The recorded data are transferred to NIES in realtime except for the cooperative stations in China at present.

To derive vertical distributions of Asian dust that can be compared directly with the dust transport models, we developed a method using the particle depolarization ratio for estimating contributions of

© 2009 IOP Publishing Ltd

dust in aerosol mixture [1,5]. The method is based on a simple assumption that the observed aerosols are an external mixture of two types of aerosols having the different aerosol depolarization ratios.

Table 1. NIES Lidar network sites (including cooperative observation sites[+])

Tsukuba (36.05N, 140.12E, 30m)	1996--	*	
[Nagasaki1 (32.78N, 129.86E, 17m)	Mar. 2001--May 2006]		
[Beijing1, China (39.9N, 116.3E)	Mar. 2001--Apr. 2007]		
[Sri Samrong, Thailand (17.15N, 99.95E)	Oct. 2001--Feb. 2005]		
Suwon[+], Korea (37.14N, 127.04E, 120m)	2002--		
[Amami-Ohshima (28.44N, 129.70E)	2002--Jun. 2005]		
[Miyakojima (24.7N, 125.3E)	2002--2005]		
Fukue (32.75N, 128.68E, 50m)	Oct. 2002--	*	**
Hefei[+], China (31.90N, 117.16E, 30m)	Oct. 2002--		**
Sapporo (43.1N, 141.3E, 30m)	2003--	*	
[Hohhot (40.94N, 111.37E)	Nov. 2003--2006]		
Toyama (36.7N, 137.1E, 28m)	Feb. 2004--	*	
Phimai, Thailand (15.18N, 102.57E, 212m)	Feb. 2005--		**
[Gosan, Korea (33.60N, 126.50E)	Mar. 2005--Jun. 2005]		
Matsue (35.21N, 133.01E)	Mar. 2005--	*	
Sendai (38.25N, 140.90E)	Apr. 2005--	*	
Cape Hedo (26.87N, 128.25E, 60m)	Aug.2005--	*	**
Seoul[+] (37.45N, 126.95E, 116m) SNU	Mar. 2006--	*	**
Shapotou[+] (37.5N, 105.0E, 1250m)	Mar. 2006--		
Nagasaki2 (32.78N, 129.86E)	Mar. 2006--	*	
Guangzhou[+] (23.16N, 113.33E, 100m)	July 2006--		
Niigata (37.84N, 138.94E, 1m)	Apr. 2007--	*	
Beijing2[+] (39.9N, 116.3E, 70m)	Apr. 2007--		
Chiba (35.65N, 140.12E, 20m)	Apr. 2007--		**
Ulaanbaatar[+] (47.92N, 106.90E, 1320m) NAMHEM	Sep. 2007--	*	
Sainshand[+] (44.87N, 110.12E, 937m) NAMHEM	Sep. 2007--	*	
Zamyn uud[+] (43.72N, 111.90E, 962m) NAMHEM	Sep. 2007--	*	

* Dust Sandstorm (DSS) Monitoring Network (ADB/GEF),** SKYNET

2. Realtime data processing system for the lidar network

Aiming at realtime validation/assimilation of chemical transport models, we developed an automatic data processing system for the lidar network. A flow chart for the system is shown in figure 1.

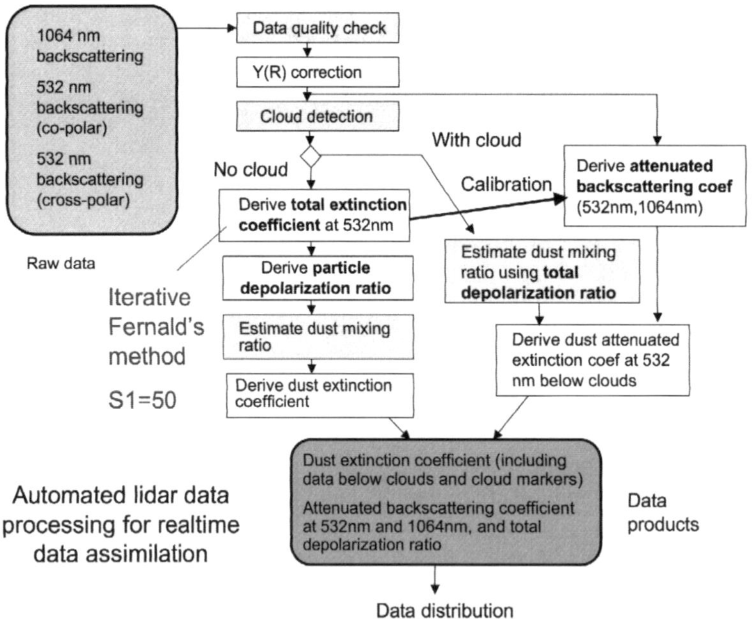

Figure 1. Flow chart for the automatic data processing system.

We apply the Fernald's method with S1=50sr iteratively with non-zero boundary value at 6 km to derive the extinction coefficient for the data without clouds, and then apply the method using the particle depolarization ratio to estimate the dust contribution [1]. For the data with clouds in the profiles below 6 km, we use the attenuated backscattering coefficient and total depolarization ratio to obtain approximate dust extinction coefficient below the clouds. We define the dust and spherical aerosol extinction coefficients as the data products for realtime validation/assimilation. We also provide the attenuated backscattering coefficients at 1064nm and 532nm, and the total depolarization ratio at 532nm for detailed analysis and more comprehensive data assimilation including various types of aerosols. The data products are updated every hour. We plan to open the numerical data through NIES web site.

3. Validation/assimilation of dust transport models using the lidar network data

Figure 2 shows the dust extinction coefficients derived from the lidar data at 11 locations with the automatic data processing system. We compare in figure 3 the results of the regional chemical transport model, Chemical Weather Forecasting System (CFORS), without data assimilation. The CFORS [6] is a multi-tracer online model build within the Regional Atmospheric Modeling System (RAMS). In the spring of 2007, the frequency of dust events was high in May. In normal years, the frequency is high in March and April. Comparing figure 2 and figure 3, it can be seen that the model without data assimilation underestimates the dust extinction coefficients in period 1 and overestimates in period 3.

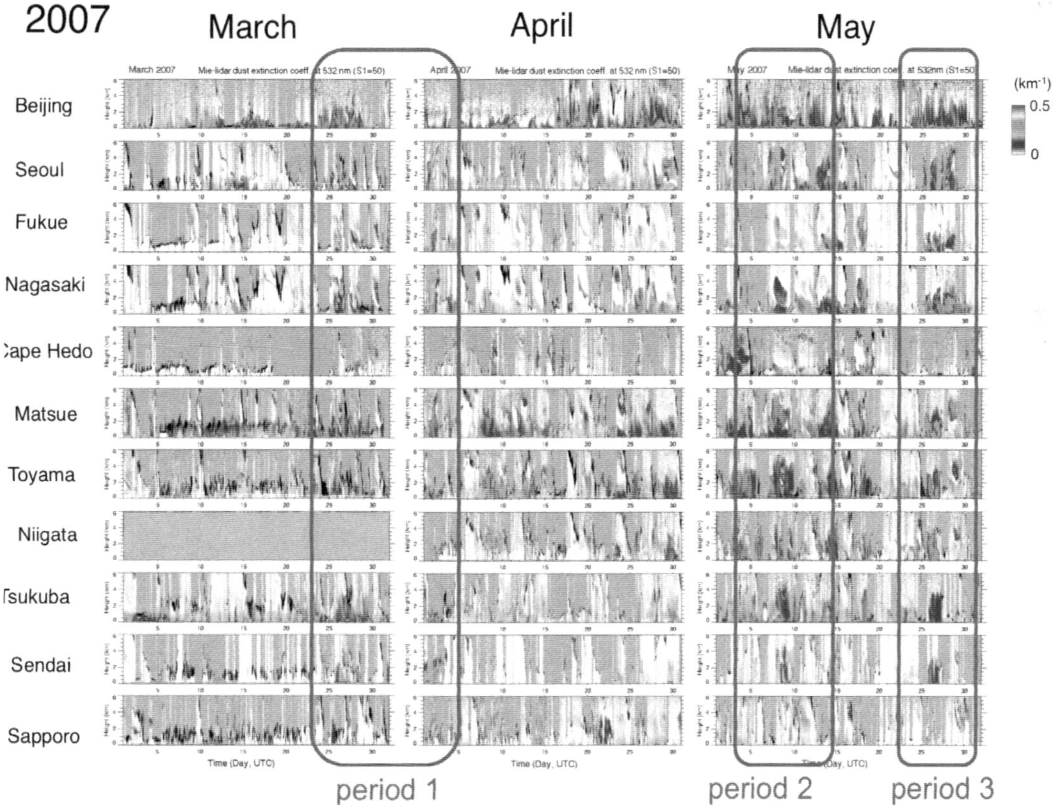

Figure 2. Dust extinction coefficients at 11 locations for March, April, and May in 2007 derived by the automatic data processing system.

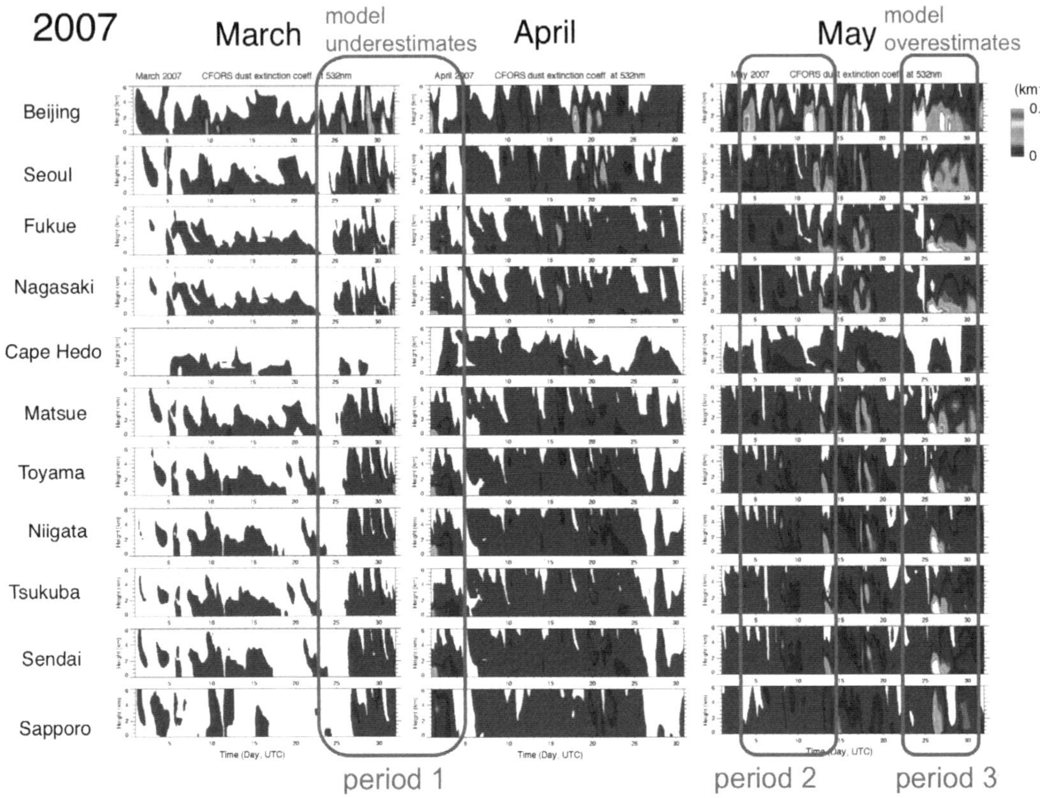

Figure 3. Dust extinction coefficients calculated with the regional dust transport model (CFORS) without data assimilation.

We developed a four-dimensional variational data assimilation system [7] and applied here to the three periods in 2007. In the data assimilation, we used a coefficient in the dust emission function as the control parameter that physically corresponds to a correction factor to the surface information. In this experiment, we used NCEP/NCAR reanalysis data as the meteorological boundary conditions. Figure 4 shows the results of the data assimilation for period 1 [8]. The dust emission was increased with the data assimilation and the resulting dust optical depth agreed better with the MODIS AOD. The analysis of the emission rate showed the emission in Mongolia was underestimated without the data assimilation. The analysis for period 2 and 3 is underway.

Figure 5 shows the comparison of the assimilated model using the lidar network data with the space-borne lidar CALIPSO [9]. We applied the forward Fernald's method to CALIPSO level 1 data and estimated dust extinction coefficient using the depolarization ratio. The vertical structure of dust layer calculated by the assimilated model agreed well with the CALIPSO observations.

One of the targets of our program is to develop a realtime data assimilation system for dust forecasting. For accurate forecasting, observations in the dust source regions are essential. We have constructed four monitoring sites including three lidars in Mongolia in a Japan International Cooperation Agency (JICA) program with National Agency for Meteorology, Hydrology and Environmental Monitoring of Mongolia (MAMHEM). We also study the lidar data processing method for dense dust cases in the source region.

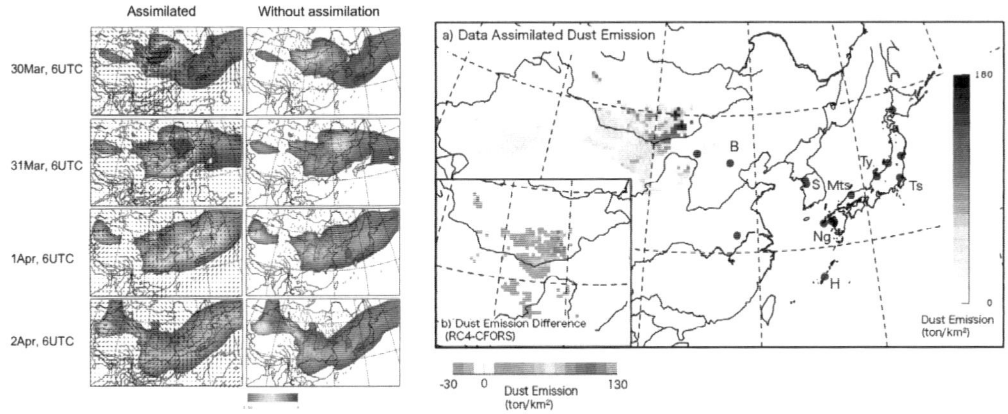

Figure 4. Dust distribution calculated by the regional dust transport model with and without data assimilation and the estimated dust emission with the assimilated model [Yumimoto et al., 2008].

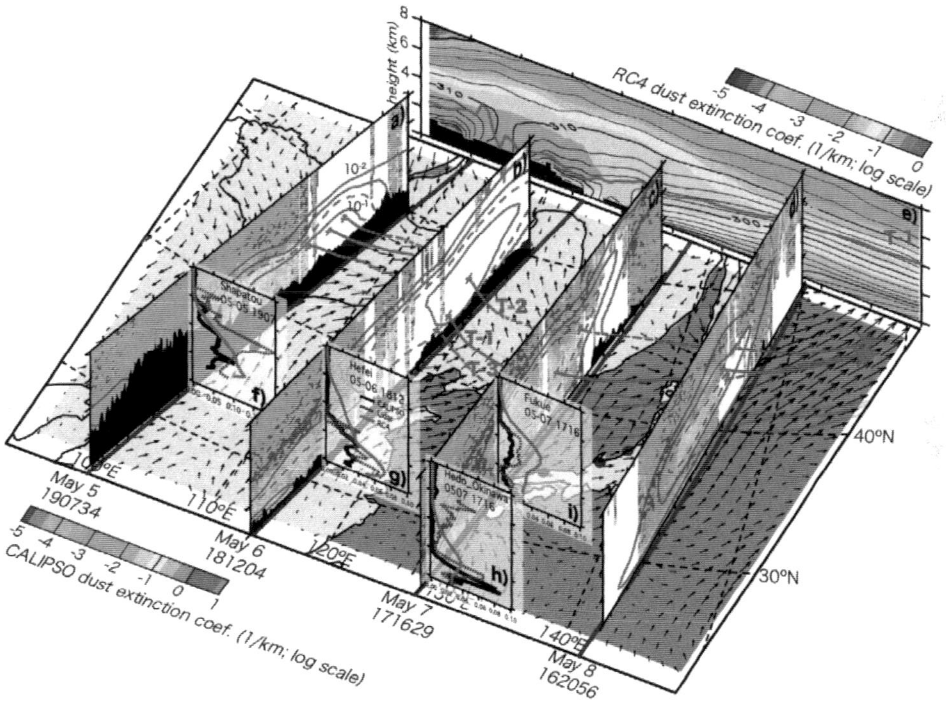

Figure 5. Comparison of the assimilated model (red contours) with CALIPSO in period 2 [Uno et al., 2008].

Acknowledgements
We thank Atsushi Shimizu, Ichiro Matsui, Tomoaki Nishizawa, Tamio Takamura, Tadahiro Hayasaka, Soon-Chang Yoon, Jun Zhou, Jixia Zhou, Yuanhang Zhang, Zifa Wang, Jugder Dulam, and all colleagues involved in the lidar network. A part of the network observation and the data assimilation study is supported by the Global Environment Research Fund from Ministry of the Environment, Japan. A part of the network observation is supported by funds from Ministry of Education, Culture, Sports, Science and Technology (MEXT), Japan on GEOSS/SKYNET.

References

[1] Shimizu, A., N. Sugimoto, I. Matsui, K. Arao, I. Uno, T. Murayama, N. Kagawa, K. Aoki, A. Uchiyama, and A. Yamazaki, 2004 Continuous observations of Asian dust and other aerosols by polarization lidar in China and Japan during ACE-Asia, *J. Geophys. Res.,* **109**, D19S17, doi:10.1029/2002JD003253

[2] Sugimoto, N., A. Shimizu, I. Matsui, U. Itsushi, K. Arao, Y. Chen, S. Zhao, J. Zhou, and C-H. Lee, 2005 Study of Dust Transport Using a Network of Continuously Operated Polarization Lidars, *Water, Air, and Soil Pollution:* **Focus 5**, 145-157.

[3] Sugimoto, N., A. Shimizu, I. Matsui et al., 2006 Network observations of Asian dust and air pollution aerosols using two-wavelength polarization Lidars, 23rd International Laser Radar Conference, July 2006 Nara, Japan (23ILRC, ISBN 4-9902916-0-3, 2006) pp. 851–854.

[4] Asian Development Bank, Prevention and Control of Dust and Sandstorm in Northeast Asia, Establishment of a Regional Monitoring and Early Warning Network for Dust and Sandstorms in Northeast Asia, Vol. 2 (Asian Development Bank, Publication Stock No. 040405, 2005, Philippines). (http://www.adb.org/Documents/Books/dust-and-sandstorm/)

[5] Sugimoto N., et al., 2003 Record Heavy Asian Dust in Beijing in 2002: Observations and Model Analysis of Recent Events, *Geophys. Res. Lett.* **30**, 12, 1640, doi:10.1029/2002GL016349.

[6] Uno, I., S. Satake, G. R. Carmichael, Y. Tang, Z. Wang, T. Takemura, N. Sugimoto, A. Shimizu, T. Murayama, T. A. Cahill, S. Cliff, M. Uematsu, S. Ohta, P. K. Quinn, and T. S. Bates, 2004 Numerical study of Asian dust transport during the springtime of 2001 simulated with the Chemical Weather Forecasting System (CFORS) model. *J. Geophys. Res.,* **109**, D19S24, doi: 10.1029/2003JD004222.

[7] Yumimoto, K., I. Uno, N. Sugimoto, A. Shimizu and S. Satake, 2007 Adjoint Inverse Modeling of Dust Emission and Transport over East Asia, 2006GL028551R, *Geophys. Res. Lett.* **34**, L08806, doi:10.1029/2006GL028551.

[8] Yumimoto, K., I. Uno, N. Sugimoto, A. Shimizu, Z. Liu, and D. M. Winker, 2008 Adjoint inversion modeling of Asian dust emission using lidar observations, *Atmos. Chem. Phys.,* **8**, 2869-2884.

[9] Uno, I., K. Yumimoto, A. Shimizu, Y. Hara, N. Sugimoto, Z. Wang, Z. Liu, and D. M. Winker, 2008 3D Structure of Asian Dust Transport revealed by CALIPSO Lidar and a 4D VAR Dust Model, *Geophys. Res. Lett.* **35**, L06803, doi:10.1029/2007GL032329.

Use of SEVIRI images and derived products in a WMO Sand and Dust Storm Warning System

M A Martínez, J Ruiz and E Cuevas

Agencia Estatal de Meteorología (AEMET), Spain

E-mail: mig@inm.es

Abstract. The Visible/IR images of SEVIRI (Spinning Enhanced Visible and Infrared Imager), on board Meteosat Second Generation (MSG) satellites, are used to monitor dust events. Satellite-based detection of dust is a difficult problem due in part to the observing-system limitations. The main difficulty is that the dust can be confused with water/ice clouds. SEVIRI is not as optimal for the viewing of dust as SEAWIFS or MODIS, due to the fact that both of them count with additional short-wavelength channels. However, the SEVIRI 15-minute loop images can detect small dust plumes as well as subtle changes from one image to the next. A description of how the AEMET, former INM, is developing the environment to support MSG satellite imagery to the WMO/GEO Sand and Dust Storm Warning System (SDS WS) for Europe, Africa and Middle East Regional Centre will be briefly presented, together with some on-going operational developments to best monitor dust events.

1. Introduction

This paper is embedded in the activities of the SDSWS for Europe, Africa and Middle East Regional Centre. The region covered by this SDSWS Centre is shown in figure 1, being mostly inside the Meteosat satellites coverage.

Figure 1. Region covered by the SDSWS

© 2009 IOP Publishing Ltd

METEOSAT-9, which belongs to the Meteosat Second Generation series, is the current nominal satellite. The main instrument on board MSG satellites is the SEVIRI radiometer. The eight thermal IR (3.9, 6.2, 7.3, 8.7, 9.7, 10.8, 12.0, 13.4 μm) and the three solar channels (0.6, 0.8, 1.6 μm) have a sampling distance of three kilometers at nadir and scan the full disk of the Earth. The high-resolution visible channel provides images with one kilometer sampling distance at nadir. Data rate limitations confine the high-resolution visible images to half the Earth disk in an east-west direction; however, the exact coverage of the Earth is programmable. Moreover, METEOSAT-8 started providing Rapid Scan Service (RSS) in May 2008 taking images every 5 minutes and covering approximately a latitude range of 15° to 69°N; this service allow monitoring dust events with unprecedented temporal resolution.

EUMETSAT is developing two MPEF day-2 aerosols related products, within its Meteorological Product Extraction Facility (MPEF). On the other hand, EUMETSAT Satellite Application Facilities for Nowcasting and Very Short Range Forecasting (NWC SAF) software package provides a dust product. This product is included in the Cloud Mask product as a flag that indicates the presence of dust. As the rest of NWC SAF products, it is obtained in near real time, at full resolution (3km x 3km at nadir) every cycle of 15 minutes for the region selected by the user. More information about this can be obtained in reference 3 or in NWC SAF web page (http://www.nwcsaf.org).

This paper is focused on the use of SEVIRI images in the monitoring of dust storm events. First, it will be shown the implementation of the RGB images recommended by EUMETSAT for monitoring desert dust with MSG. Later, some early developments made at the AEMET to enhance the analysis of dust information contained in the SEVIRI VIS channels will be presented.

2. Continuous monitoring of desert dust using EUMETSAT RGB images

EUMETSAT recommends the use of some RGB images for dust monitoring in the MSG interpretation guide which is available on its web page (http://oiswww.eumetsat.org/WEBOPS/msg_interpretation/index.html). The Dust RGB is an RGB composition based upon infrared channel data from SEVIRI (table 1) where dusty pixels appear in pink colors. It is designed to monitor the evolution of dust storms over deserts during both day and night. Dust RGB images are available free of charge every hour through the EUMETSAT web page: http://www.eumetsat.int/Home/Main/Image_Gallery/Real_Time_Imagery/index.htm.

Table 1. Dust RGB composition

Dust (night & day)					
RGB colour plane	Channel (difference)	MIN	MAX	GAMMA	Prominent features
R	12.0 - 10.8	-4 K	+2 K	1.0	Dust(over land)
G	10.8 - 8.7	0 K	+15 K	2.5	Thin Cirrus
B	10.8	261 K	289 K	1.0	Contrails

The Natural RGB is an RGB composition based upon visible channels data from SEVIRI (table 2). It is recommended for the tracking of dust across the oceans during the day.

Table 2. Dust RGB composition

Natural Colours					
RGB colour plane	Channel	MIN	MAX	GAMMA	Prominent features
R	1.6	0 %	100 %	1.0	Ice/water clouds, fog, snow
G	0.8	0 %	100 %	1.0	Dust(over ocean),smoke
B	0.6	0 %	100 %	1.0	(Green) vegetation

The strongest sand storm monitored with MSG satellite took place in March 2004 and it has been selected as a clear example. The intensity and the wide extension affected this dust event can be appreciated in the Dust RGB image shown in figure 2 (3rd March at 12Z). The dust event affected the whole Sahara desert and crossed the Atlantic Ocean in the following days. Loops of dust RGB images for this event are available on EUMETSAT and on WMO WS web page (http://salam.upc.es/wmo). On the other hand, the dust over ocean can be seen in the Natural RGB image shown in figure 3 (5th March at 9Z).

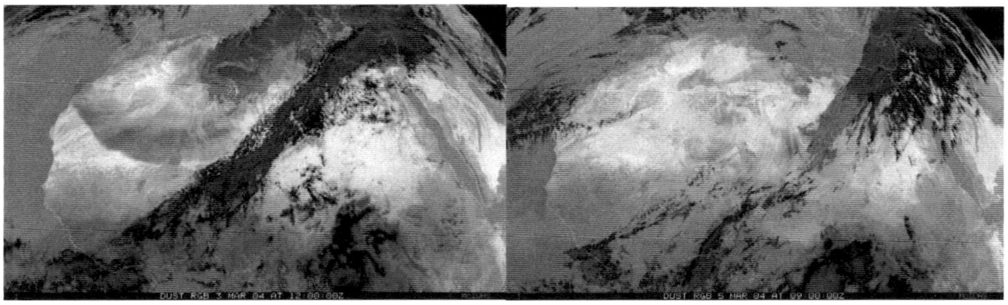

Figure 2. Dust RGB images during March 2004 dust event. 3rd March 2004 12Z (left) and 5th March 2004 9Z (right)

3. AEMET development (The current Approach)

3.1. Normalized natural RGB

Due to the fact that the incoming radiance is different depending on the position of the earth throughout the year, and also depending on the moment the satellite makes the observation. It will be convenient to normalize in a certain manner the radiance values obtained by the radiometer in order to compare situations that can last several days and that can appear at any time during the year.

The Bidirectional Reflectance Factor (BRF) (equation 1) allows to correct the visible radiances (RAD) taking into account the solar zenith angle (SZA), the Sun Earth distance (SE), the channel bandwidth (BW) and the Sun Irradiance (SI).

$$BRF = \frac{\pi \cdot RAD \cdot BW \cdot SE^2}{SI \cdot \cos(SZA)} \tag{1}$$

The BRF variable factors can be grouped in one term, called normalized Radiance in this paper (equation 2):

$$RADn = \frac{RAD \cdot SE^2}{\cos(SZA)} \tag{2}$$

The normalized radiance for the 0.6, 0.8 and 1.6 μm SEVIRI channels is routinely calculated using a command programmed in the McIDAS (Man computer Interactive Data Access System) environment (http://www.ssec.wisc.edu/mcidas). This command has been developed at the AEMET. The coefficients for the calculation of SZA and SE have been obtained from [1]. The Natural RGB command is applied to the normalized radiances and the obtained RGB images are called normalized Natural RGB images.

In figure 3, the Natural RGB image (left) is compared with the normalized Natural RGB image (right) for the same slot. The normalized Natural RGB shows a better dust monitoring, to simple sight.

It also allows discriminating dust during more hours (sunrise to sunset). Moreover, it is possible to compare normalized Natural RGB images throughout the year.

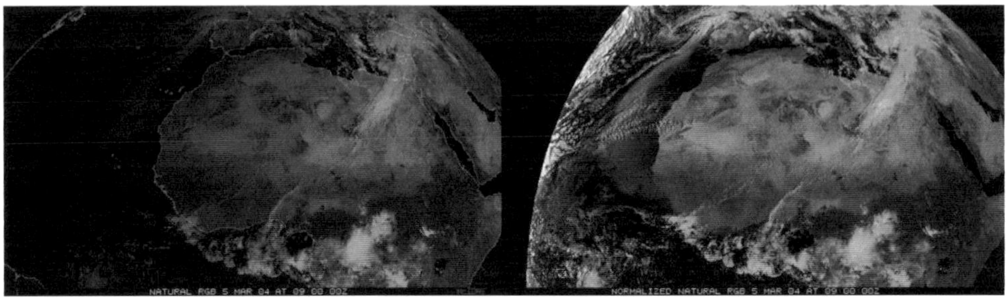

Figure 3. Natural RGB images during March 2004 dust event. Not normalized on the left and normalized on the right (5 March 2004 9Z)

3.2. Dust monitoring over land pixels using SEVIRI visible channels

Over land pixels the visible channels are not so efficient in the dust monitoring. The 19th-25th February 2007 sand storm event has been selected to investigate the information contained in the visible channels over land pixels. The image of NWC SAF dust flag for the 21st February 2007 at 12 Z is shown in figure 4, where pixels in green represent zones where the algorithm has detected the presence of dust.

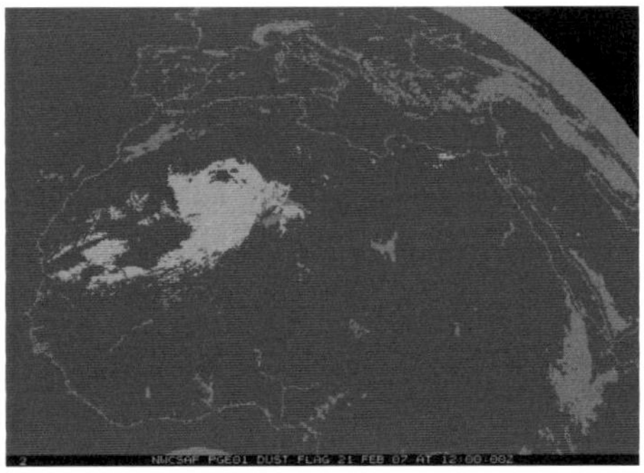

Figure 4. Dust flag of SAFNWC Cloud Mask (21 February 2007 at 12Z). Dusty pixels appear green

In figure 5, the Dust RGB image (on the left), with dusty pixels in pink color, is shown together with the normalized radiances of each visible channel (on the right) for the same date and time. On the visible channels over the region of interest (red square), it can be seen how the presence of the dust is mixed with the ground information.

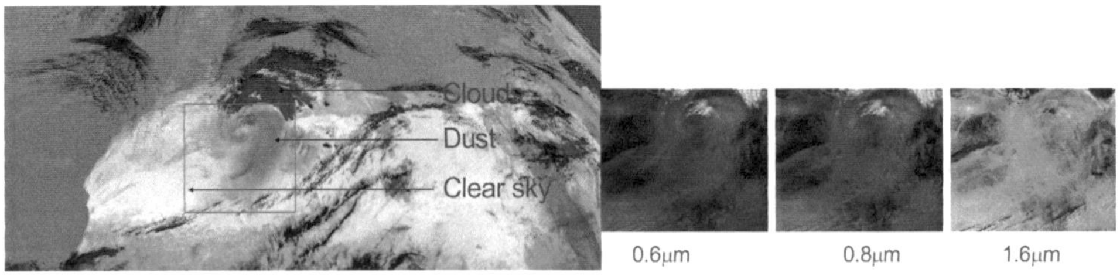

Figure 5. Dust RGB and 0.6, 0.8 and 1.6 Normalized Radiance inside the red square region (21 February 2007 12Z)

The objective is to detect the presence of dust in the visible SEVIRI images. In order to reach it, it is necessary to detach the information of the background introduced by the textures of the desert, and to enhance the information coming from the dust. This task has been performed building monthly cloudless radiance masks at 12 Z for each channel (0.6, 0.8, 1.6 μm). The used criteria to construct the monthly mask for each channel are different for land and sea pixels. Over every land pixel, the array containing the radiances for the month is sorted (in ascending order). Later on, the day with the highest NDVI is searched among the third part of the data with lower values (excluding the minimum value). In the case of sea pixels, all the data for the month is managed together and the value with the maximum frequency of the histogram for the month is selected. All the sea pixels are set to this fixed value.

The normalized Natural RGB image for the dust event of the 21st February is shown in figure 6 (right); the dust appears as a shadow in the image. When the Natural RGB is applied to the masks, the ground information is retained by the monthly mask. As an example the February 2007 RGB mask image is shown in figure 6 (left).

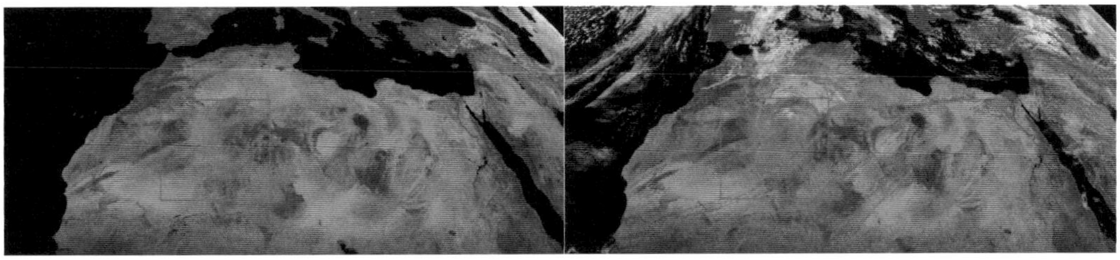

Figure 6. Normalized Natural RGBs: monthly mask (left) and 21 February 2007at 12Z (right)

3.3. Prototype of dust classifier

Detection of dust over bright land such as deserts is more difficult than over ocean, and it requires IR information to separate the dust signal from the other components of the scene. In the Dust RGB image of figure 7 (left) it can be seen some variations in the dust intensity (different pink colour intensity). Four classes have been established (figure 7, right) following the physical basis of the Dust RGB: high intensity (white), medium intensity (yellow), low intensity (red) and clear-sky (blue). Table 3 contains the empirical thresholds used to separate the classes.

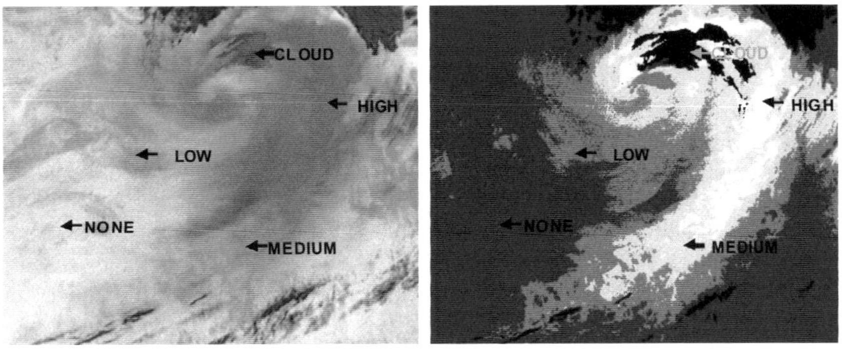

Figure 7. Dust RGB 21st February 2007 12Z (left) and classification of dust intensity (right)

The main issue is to study the dust signal on the visible channels. The BRF scatter-plot of each visible channel versus the corresponding BRF monthly mask for each class have been analyzed; in figure 8, the VIS0.6 scatter-plots are shown as an example.

Table 3. Criteria used to obtain the dust intensity classification.

	HIGH	MEDIUM	LOW	CLOUD	NONE
IR12.0-IR10.8	> 3 K	> 1.9 K	> 1 K	--	Other case
IR10.8-IR8.7	< 2 K	< 4 K	< 7 K	--	Other case
IR10.8	--	--	--	< 275 K	Other case

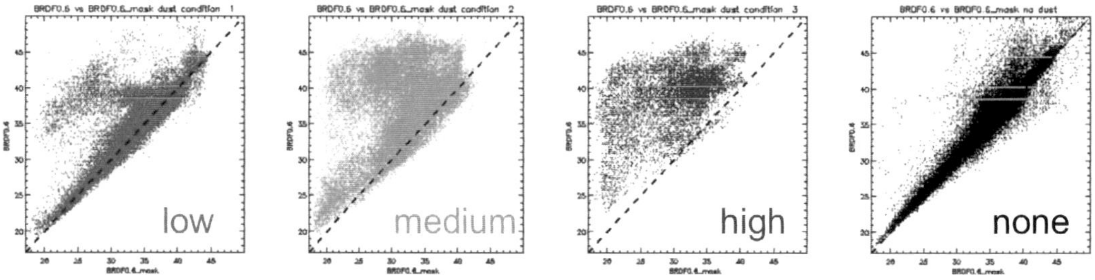

Figure 8. Scatter-plot of BRF VIS0.6 versus the BRF VIS0.6 monthly mask for each class of dust intensity.

Differences between the image and its monthly mask can been detected in all channels for medium and high intensity classes. As the differences between the image and the mask are small, the use of the logarithm of the difference is more adequate to enhance the presence of dust. In figure 9, it is shown an RGB composition with the logarithm of the differences between each BRF and its mask (following the same Natural RGB channel order). It can be seen how the textures of the desert have disappeared and how the regions affected by the presence of dust are clearly enhanced.

Figure 9. RGB composition with the logarithm of the differences between each BRF and its mask (21Ffebruary 2007 12Z).

4. Conclusions

The SEVIRI spectral range, its spatial resolution (3 km in the IR bands at nadir), and its 15 minutes cycle allow to better monitor desert dust events. The RGB images constitute an easy and fast approach for dust monitoring. The used of visible normalized radiances improve the dust monitoring when the Natural RGB is applied. The thermal infrared thresholds used in the preliminary classification have been obtained empirically and could be improved in future works. An operational environment system for dust monitoring based on web pages is being developed at the AEMET. However, further work is needed to improve the quality of the dust estimation using visible channels over land pixels.

Satellite data can complement dust models and in situ measurements to allow a better monitoring of dust events. In particular, the comparison with DREAM numerical model outputs (http://www.bsc.es/projects/earthscience/DREAM) is included in the SDSWS project.

References

[1] Liou, K.N., 2002 An introduction to Atmospheric Radiation. Second Edition. International Geophysics Series. Volume 84. Academic Press. Pages 44-50.

IOP Conf. Series: Earth and Environmental Science 7 (2009) 012005 doi:10.1088/1755-1307/7/1/012005

Asian Dust Particles Impacts on Air Quality and Radiative Forcing over Korea

Y J Kim[1], Y M Noh[1], C H Song[1], S C Yoon[2] and J S Han[3]

[1] Advanced Environmental Monitoring Research Center (ADEMRC). Department of Environmental Science and Engineering. Gwangju Institute of Science and Technology (GIST), Gwangju, Korea
[2] School of Earth and Environment Sciences, Seoul National University, Seoul, Korea
[3] National Institute of Environmental Research (NIER), Ministry of Environment, Incheon, Korea

Email: yjkim@gist.ac.kr

Abstract. Asian Dust particles originated from the deserts and loess areas of the Asian continent are often transported over Korea, Japan, and the North Pacific Ocean during spring season. Major air mass pathway of Asian dust storm to Korea is from either north-western Chinese desert regions or north-eastern Chinese sandy areas. The local atmospheric environment condition in Korea is greatly impacted by Asian dust particles transported by prevailing westerly wind. Since these Asian dust particles pass through heavily populated urban and industrial areas in China before it reach Korean peninsular, their physical, chemical and optical properties vary depending on the atmospheric conditions and air mass pathway characteristics. An integrated system approach has been adopted at the Advanced Environment Monitoring Research Center (ADEMRC), Gwangju Institute Science and Technology (GIST), Korea for effective monitoring of atmospheric aerosols utilizing various in-situ and optical remote sensing methods, which include a multi-channel Raman LIDAR system, sunphotometer, satellite, and in-situ instruments. Results from recent studies on impacts of Asian dust particles on local air quality and radiative forcing over Korea are summarized here.

1. Introduction

Asian dusts that sometimes blanket East Asia have serious impacts on the global climate system. These dust particles originated from the Taklimakan or the Gobi Deserts of China and Mongolia travel over Korea most frequently in late winter and early spring. The dust particles not only reflect the incoming solar radiation back to the space like other atmospheric aerosols, but also affect cloud properties by changing the number concentration of cloud droplets, which in turn alter both cloud optical depth and cloud lifetime [1]. In addition, the particles can affect the radiative heating structure of the atmosphere, thereby changing the clouds and atmospheric general circulation.

At present, tropospheric aerosols pose one of the largest uncertainties in model calculations of the climate forcing due to man-made changes in the composition of the atmosphere [2]. The International Global Atmospheric chemistry Program (IGAC) had organized a series of Aerosol Characterization Experiments (ACE) that integrated in situ measurements, satellite observations, and models to reduce the uncertainty in calculation of the climate forcing due to aerosol particles. ACE-Asia was the fourth in this series of experiments, which included an intensive field study in northeast Asia during the

© 2009 IOP Publishing Ltd

spring of 2001 [3]. ACE-Asia focuses on aerosol outflow from Asia to the Pacific basin since both anthropogenic aerosols and mineral dust from the Asian continent greatly affect the atmospheric environment and radiation balance in the downwind regions. Since ACE-Asia, an integrated system approach as shown in figure 1 has been adopted at the Advanced Environment Monitoring Research Center (ADEMRC), Gwangju Institute Science and Technology (GIST), Korea for effective monitoring of atmospheric aerosols utilizing various in-situ and optical remote sensing methods, which include a multi-channel Raman LIDAR system, sunphotometer, satellite, and in-situ instruments. Based on the measurement results made with the integrated monitoring system at GIST/ADEMRC Asian dust particles' effects on air quality and radiative forcing are summarized in this report.

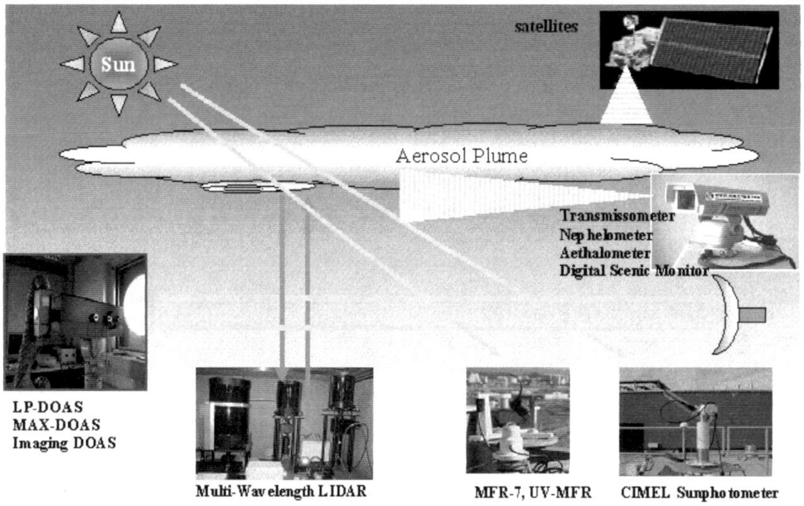

Figure 1. GIST integrated monitoring system for atmospheric aerosol.

2. Asian Dust Particles

Asian dust particles that originated from the deserts and losses area of the Asian continent are often transported over Korea, Japan, and the North Pacific Ocean during the spring season. During the last decades in East Asia region, the occurrence number of Asian dust days continuously increased. Occurrence number of Asian dust days in Seoul during the period from 1980 to 2007 is plotted in figure 2, which shows large increase since 2000.

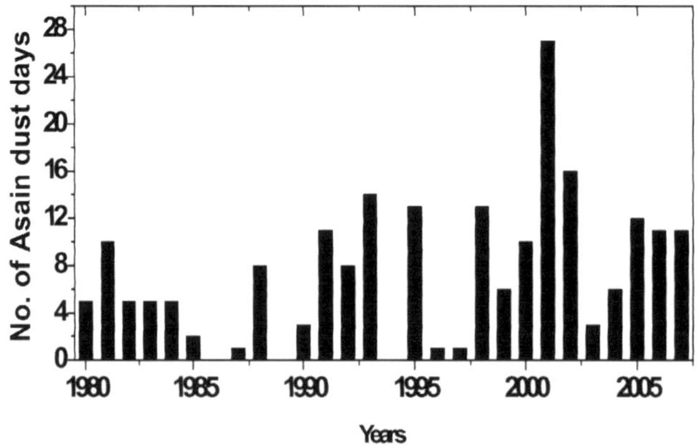

Figure 2. Occurrence number of Asian dust days in Seoul during the period from 1980 to 2007.

Asian dust particles consisted of many particles of different origins externally mixed to form loose conglomerates and some particles of internal mixture of various compounds. [4] summarized the characteristics of Asian dust particles observed during the ACE-Asia. Dust particles had pervasive influence on chemical composition, size distribution, and optical properties of the aerosol. Coarse-mode dust particles were less absorbing and less hygroscopic than the pollution particles. There was little correlation in light scattering and absorption by the sub- vs. super-micrometer particles. It was found that dust particles were commonly mixed with pollutants below ~ 2km, influencing on the optical properties of the submicrometer particles. The mixing of black carbon (BC) with dust was commonly detected during the period, but only certain types of BC particles were aggregated.

Since much more coal and biomass are burned in Northeast Asia, Asian aerosol sources unlike those in Europe and North America add more absorbing soot and organic aerosol to atmosphere. Mixing state and morphology of dust in outflow were studied during the ACE-Asia IOP. Aggregation of black Carbon (BC) particles with dust was common in Northeast Asian dust outflow periods, even at high altitudes. There was a variety of BC types which have been observed to be aggregated with dust particles as shown in figure 3. It shows soot particles stuck to a mineral dust particle collected on a polycarbonate Nuclepore filter [5]. It also shows that BC aggregated with dust is different in size and shape from normal BC. Dust particles externally mixed with pollution particles during long-range transported showed increased light-absorbing characteristics. [6] Reported that the single-scattering albedo was observed to be 0.93, 0.90, and 0.84 during three Asian dust event periods over Gwangju in 2001. The difference of single-scattering albedo was affected from pollutions on the dust particles during long-range transport. [7] reported that cases where aerosol optical depth was predominately coarse mode dust aerosol in the spring of 2001, the absorption was greater in eastern Asia compared to the source regions, with ~ 0.93 of single-scattering albedo at Dunhuang, China (near to the major Taklimakan dust source), which was 0.04 higher than that at Beijing. Possible reasons for differences in dust absorption magnitude include interactions between dust and fine mode pollution aerosol and also variability of dust optical properties from different source regions in China and Mongolia. Biomass burning product is one of the main substances that mixes with Asian dust. In Northeastern Asia, agricultural wastes are commonly burned in open fields soon after harvest, and these burning operations are a major source for anthropogenic aerosols and gases [8].

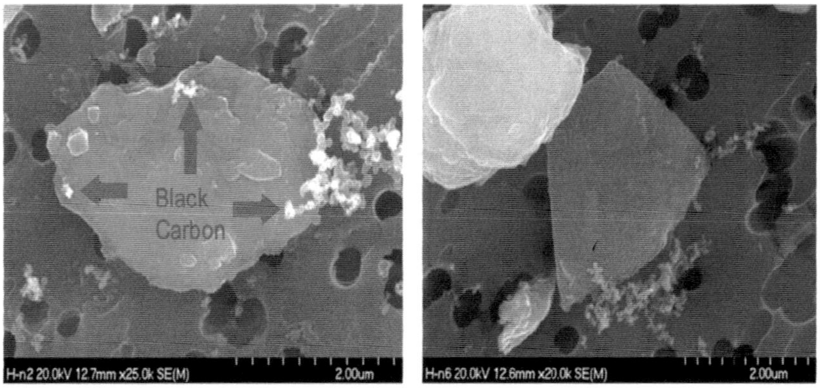

Figure 3. Scanning electron microscope (SEM) images of mineral dust particles measured at Gwangju, Korea in 2001 [5].

Optical characteristics of Asian dust was studied in 2004 and 2005 at Anmyoen Island (36.32N, 126.19E) and Gwangju (35.10N, 126.53 E), Korea using the GIST/ADEMRC (Gwangju Institute of Science & Technology/ADvanced Environmental Monitoring Research Center) multi-wavelength Raman lidar system [9]. Asian dust particles were separated from other types particles by high depolarization ratio (< 0.15) and low Angstrom exponent. Also, lidar ratio of Asian dust particles was distinguished from other types aerosols (e.g., smoke, haze, local pollution particles) by high light absorbing characteristics at UV wavelength than visible wavelength.

Table 1. Optical parameters measured by Raman Lidar [9].

Aerosol Type	**Date**	**α** [1]	**δ** [2]	**S_a**	
				532 nm	**355 nm**
Asian Dust	March 15th, April 2nd (L and H), 8th, 12th, 15th, 2004	0.78 ± 0.20	0.26 ± 0.10	47 ± 7	49 ± 12
Forest Fire Smoke	April 3rd 2004, May 24th, 2005	1.01 ± 0.24	0.08 ± 0.01	65 ± 6	52 ± 10
Haze from China	May 27th 2005	1.78	0.08	70	64
Local Haze	May 28th, 29th, 30th, 31st, June 1st, 4th, 5th, 7th, 8th, 2005	1.45 ± 0.30	0.08 ± 0.02	82 ± 7	73 ± 7

[1]Angstrom Exponent [2]Depolarization ratio

Table 1 summarizes the observation results. Lidar ratio of dust particles was higher at 355 nm (56 ± 10 sr) than at 532 nm (51 ± 6 sr) over Gwangju. Similar spectral behavior was also measured at Anmyeon with 47 ± 7 and 45 ± 5 sr at 355 and 532 nm, respectively. Average lidar ratio value, 51 sr observed at 532 nm over Gwangju for Asian dust particles in spring was lower than those for non-dust (60 sr) and smoke (65 sr) aerosol particles. Lower lidar ratio of dust particles (45 sr at 532 nm) was also measured at Anmyeon, compared to that of anthropogenic particles (79 sr). The spectral dependence of the lidar ratio can be utilized to distinguish aerosol types in remote sensing. Large differences in lidar ratio were detected between long-range transported dust particles and accumulation mode pollution particles mixed with biomass burning aerosol. Figure 4 shows relationship among lidar ratio (S_a), depolarization ratio (δ), and Angstrom Exponent (α) according to different aerosol types. Asian dust particles were distinguished from other types aerosols by high depolarization ratio, low lidar ratio and Angstrom exponent at 355 and 532 nm.

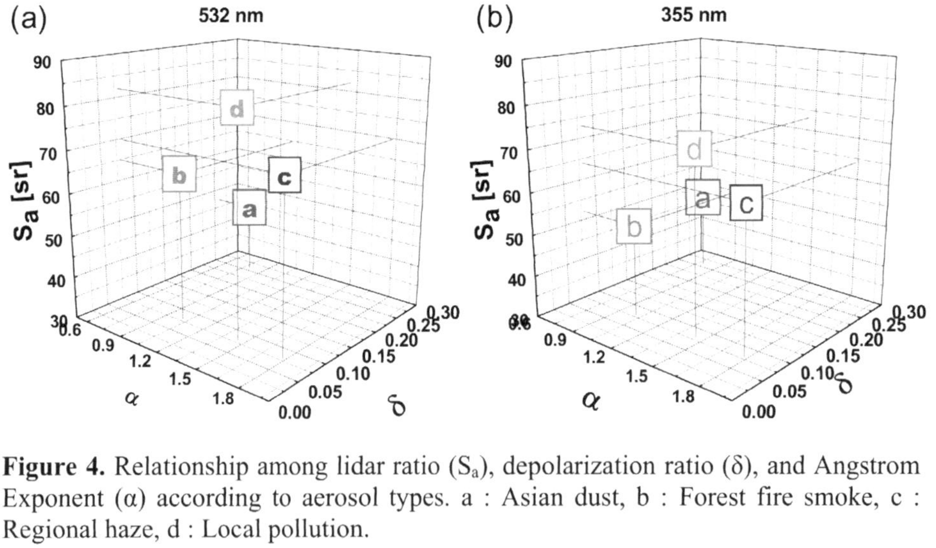

Figure 4. Relationship among lidar ratio (S_a), depolarization ratio (δ), and Angstrom Exponent (α) according to aerosol types. a : Asian dust, b : Forest fire smoke, c : Regional haze, d : Local pollution.

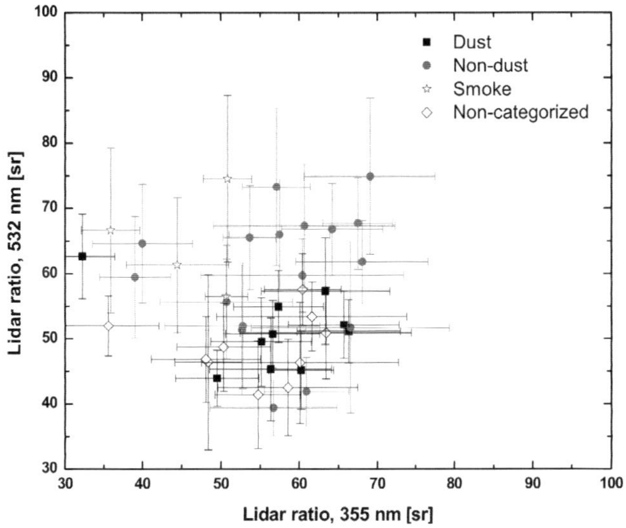

Figure 5. Lidar ratio at 355 nm vs. 532 nm for different aerosol types observed in spring 2004 and 2005 at Gwangju, Korea [10].

Figure 5 shows the lidar ratio distribution of each categorized aerosol at the wavelengths of 355 and 532 nm observed in spring of 2004 and 2005 at Gwangju, Korea. Dust and smoke aerosols can be clearly distinguished in figure 5. However, non-dust aerosol shows wide variations in the lidar ratio at both wavelengths compared to those of the other two types of aerosols.

3. Radiative Effect of Asian Dust

In order to quantify the direct radiative effect of the combined natural and anthropogenic aerosol during the ACE-Asia IOP, optical properties of atmospheric aerosols have been analyzed. The Asian Dust particle was perturbed by the mixture of both continental and regional aerosols, thus, single scattering albedo was measured to be relatively low. High loading of Asian Dust particles resulted in forming of the externally mixed dust particles with agglomerated BC particles. Mass fraction of BC was averaged to be 0.27 during three Asian Dust events. Its uncertainty bound was calculated to be 0.12 – 0.25. In addition, it was revealed that the quantification of coarse mass scattering efficiency should be investigated chemically as well as optically to improve the chemical transport models and to refine radiative transfer models. The multi-year AERONET sun/sky radiometer measurements, performed at Gosan (a super-site of the ACE-Asia 2001 campaign and of the ABC-EAREX international long-term sustainable experiment since 2005) and Anmyon (Korea Global Atmosphere Watch Center, National Institute of Meteorological Research) in Korea, have been analyzed to investigate the averaged climatological values of the primary optical and physical properties of spring Asian dust particles and to compare the column-integrated aerosol properties and radiative forcing during the Asian dust periods with those during the non-Asian ones [11]. Figure 6 shows the instantaneous aerosol radiative forcing (ΔF) at the surface and TOA for the five periods (Asian dust: SPR-AD, Non-Asian dust overall: SPR-NAD, Dust free: SPR-NAD-F, Elevated dust layer: SPR-NAD-C, May-June non-Asian dust: MJ-NAD). Since ΔF is directly proportional to the column aerosol amount (i.e., represented as t), the largest ΔF both at the surface and TOA was apparent during SPR-AD both at Gosan and Anmyon, with about $-75 Wm^{-2}$ at the surface and $-37 Wm^{-2}$ at TOA. MJ-NAD showed the second largest values of ΔF, but the difference for other periods was not significant, because the aerosol properties were not much different.

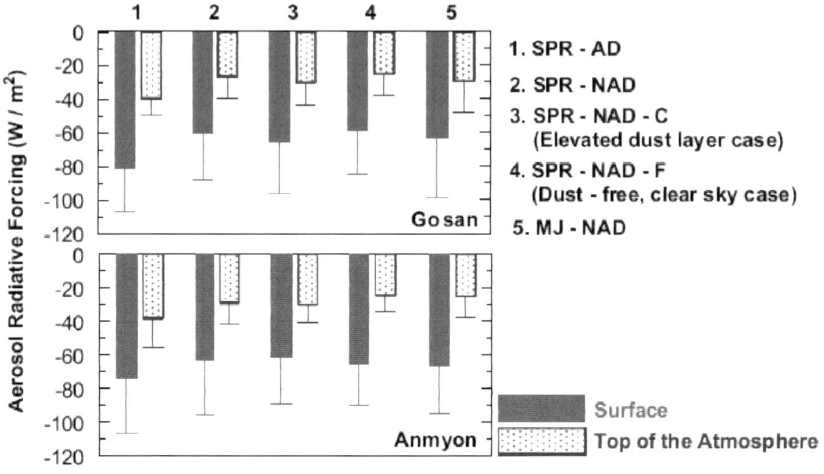

Figure 6. Aerosol-direct radiative forcing both at the surface and top of the atmosphere during the spring Asian dust period (SPR-AD), spring non-Asian dust (SPR-NAD, SPR-NAD-C, SPR-NAD-F) and May–June non-Asian dust (MJ-NAD) periods [11].

Figure 7 presents the model calculated spatial distributions of the column concentrations of five major particulate constituents of PM2.5: (i) nitrate (NO_3^-), (ii) sulfate (SO_4^{2-}), (iii) dust, (iv) BC, and (v) SOAs [12]. This model results can be used to predict future atmospheric variation in East Asia by comparison with observation results.

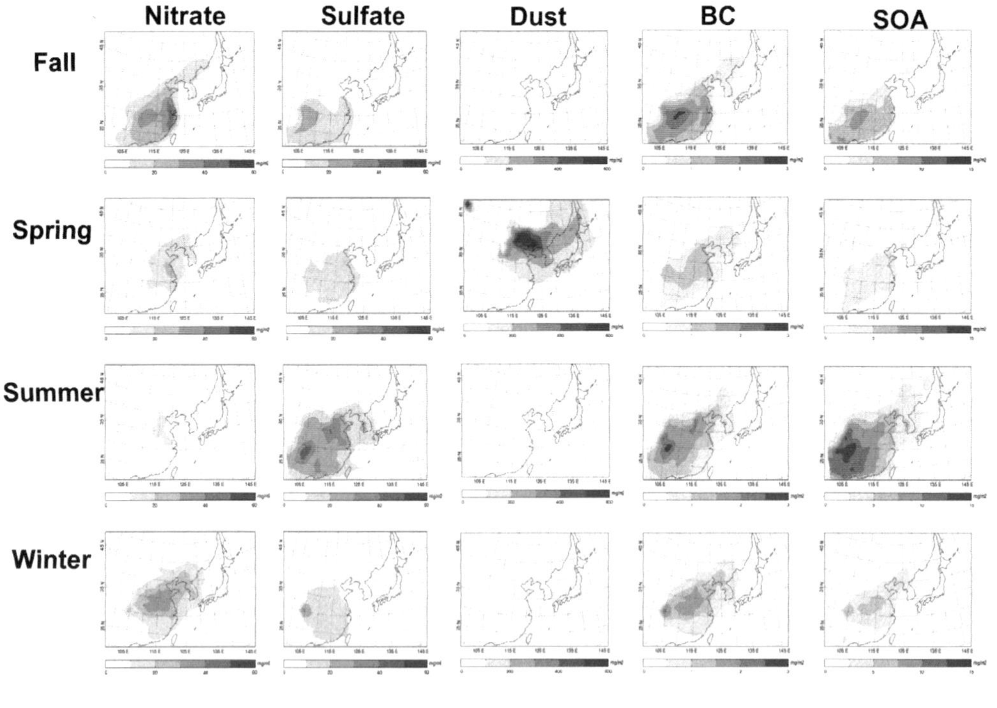

Figure 7. Spatial distributions of the column concentrations of five major particulate constituents over Northeast Asia between 2001 and 2003 [12].

4. Conclusion

The physical, chemical, and optical properties of Asian dust were studied by integrated monitoring approach and radiative forcing by them was estimated. Aerosol concentrations and properties observed in East Asian were strongly affected by the occurrence of Asian dust storm in spring. Aggregation of BC with dust was common in East under the Asia dust outflow conditions. Light-absorbing pollutants typically are added to dust plumes during transport. Integrated monitoring system with LIDAR, sunphotometer, and satellite can provide important information on aerosol and can be used for model validation and improvement. It can also permit us to obtain more insight in the mechanisms of long-range transport of dust particles, and their impact on air quality in Korea. Furthermore, optical and microphysical parameters retrieved from a multi-wavelength Raman lidar data allow us to determine the light-absorbing characteristics as well as the vertical distribution of atmospheric aerosol. It has been shown that additional measurement capability of a multi-wavelength Raman lidar system can be utilize for aerosol type characterization. Since ACE-Asia, continuous monitoring of atmospheric aerosol and atmospheric environment change over Korea has been conducted through the intensive measurements of physical, chemical, and optical properties of atmospheric aerosol through various intensive studies including international projects such as ABC (Atmospheric Brown Cloud) project and ADOES (Asian Dust and Ocean Ecosystem)/ SOLAS program.

Acknowledgement
This work was supported by the Korea Meteorological Administration Research and Development Program under Grant CATER 2007-4108.

References

[1] Ackerman, A. S., O. B. Toon, D. E. Stevens, A. J. Heymsfield, V. Ramanathan, and E. J. Welton, 2000 Reduction of tropical cloudiness by soot. *Science*, **288**, 1042– 1047.
[2] Intergovernmental Panel on Climate Change (IPCC), Climate Change 2007: The Scientific Basis, edited by J. T. Houghton et al., Cambridge Univ. Press, New York. 2007.
[3] Huebert, B. J., T. Bates, P. B. Russell, G. Shi, Y. J. Kim, K. Kawamura, G. Carmichael, T. Nakajima, 2003 An overview of ACEAsia: strategies for quantifying the relationships between Asian aerosols and their climatic impacts. *J. Geophys. Res.*, **108** (D23), 8633. doi:10.1029/2003JD003550.
[4] Arimoto, R., Y. J. Kim, Y. P. Kim, P. K. Quinn, T. S. Bates, T. L. Anderson, S. Gong, I. Uno, M. Chin, B. J. Huebert, A. D. Clarke, Y. Shinozuka, R. J. Weber, J. R. Anderson, S. A. Guazzotti, R. C. Sullivan, D. A. Sodeman, K. A. Prather, I. N. Sokolik, 2006 Characterization of Asian Dust during ACE-Asia, *Global and Planetary Change,* **52**, 23–56.
[5] Kim, K. W., Z. He, and Y. J. Kim, 2004 Physico-chemical characteristics and radiative properties of Asian dust particles observed at Kwangju, Korea, during the 2001 ACE-Asia IOP. *J. Geophys. Res.,* **109**, D19S02, doi:10.1029/2003JD003693.
[6] Kim, D. H., Z. He, Y. J. Kim, 2004. Aerosol optical properties over east Asia determined from ground-based sky radiation measurements. *J. Geophys. Res.,* **109**, D02209, doi:10.1029/2003JD003387.
[7] Eck, T. F., B. N. Holben, O. Dubovik, A. Goloub, H. B. Chen, B. Chatenet, L. Gomes, X. Y. Zhang, S. C. Tsay, Q. Ji, D. Giles, I. Slutsker, 2005 Columnar aerosol optical properties at AERONET sites in central eastern Asia and aerosol transport to the tropical mid-Pacific. *J. Geophys. Res.,* 110, D06202, doi:10.1029/2004JD005274.
[8] Ryu, S. Y., J. E. Kim, Z. He, Y. J. Kim, 2004 Chemical Composition of Post-Harvest Biomass Burning Aerosols in Gwangju, Korea. *Journal of the Air & Waste Management Association* **54** (9), 1124–1137.
[9] Noh, Y. M., Y. J. Kim, B. C. Choi, T. Murayama, 2007 Aerosol Lidar Ratio Characteristics Measured by a Multi-Wavelength Raman Lidar System at Anmyeon Island, Korea, *Atmospheric Research*, doi:10.1016/j.atmosres.2007.03.006

[10] Noh, Y. M.. Y. J. Kim, and D. Muller, Seasonal characterisets of lidar ratio measured with a Raman Liar at Gwangju, Korea in spring and autumn, Atmospheric Environment, **42**(9), 2208-2224 (March 2008)

[11] Kim, S. W., S. C. Yoon, J. Y. Kim. 2008 Columnar Asian dust particle properties observed by sun/sky radiometers from 2000 to 2006 in Korea. *Atmospheric Environment,* **42** (2008) 492–504.

[12] Song, C. H., M. E. Park, H. J. Ahn, K. H. Lee, Y. Lee, Y. J. Kim, K. M. Han, J. Kim, Y. S. Ghim, and Y. J. Kim, 2008 An investigation into seasonal and regional aerosol characteristics in East Asia using model-predicted and remotely-sensed aerosol properties. *Atmsopehric Chemistry & Physics*, 2008,6627-6654,2008.

IOP Conf. Series: Earth and Environmental Science 7 (2009) 012006 doi:10.1088/1755-1307/7/1/012006

Modelling mineral dust emissions

B Laurent[1], B Marticorena[2], G Bergametti[2], I Tegen[1], K Schepanski[1] and B Heinold[1]

[1] Leibniz-Institut für Troposphärenforschung, IfT, Permoserstraße 15, 04318, Leipzig, Deutschland.
[2] Laboratoire Inter-universitaire des Systèmes Atmosphériques, LISA, UMR 7583 CNRS, Universités Paris VII-XII, 61 Avenue du Général de Gaulle, 94010, Créteil, France.

Email: laurent@tropos.de

Abstract. The biogeochemical cycle of mineral dust is of major interest to understand climatic changes. Moreover, these particles can also cause risks for human health and societal activities in regions in the neighbourhoods of arid and semi-arid source areas of dust emission. To estimate and forecast atmospheric dust concentrations and their impacts, the correct description of the spatial and temporal variability of dust emission occurrences and intensities is a prerequisite. The explicit dust emission models provide a physical description of the main processes involved in dust production. They allow describing the spatio-temporal variability of the non-linear phenomenon of dust production if their input parameters (surface and soil features, surface winds) are accurately described. The recent developments, the current limits of these emission models, and some of their applications using relevant surface, soil and meteorological databases to simulate dust emissions are presented here.

1. Introduction

Atmospheric mineral dust is mainly produced by aeolian erosion acting in arid and semi-arid areas. Mineral aerosols have an impact on the Earth's radiative budget by absorbing and scattering incoming solar and outgoing terrestrial radiation (e.g. [1]). The evaluation of this impact still represents a major uncertainty on the understanding of climatic changes. Mineral dust is involved in heterogeneous and multiphase atmospheric chemistry, affecting photo-oxidant concentrations and the composition of precipitation [2,3]. It also contributes to the biogeochemical cycles of many elements, as Fe and P, suspected to be limiting in isolated ecosystems (e.g. open oceans, the Amazon forest) [4,5,6]. At a regional scale, emitted mineral dust represents risks and nuisances for the exposed populations living close to the source areas.

To evaluate the mineral dust impacts, their concentration fields have to be determined precisely. Modelling dust emissions is of key importance to reproduce dust atmospheric concentrations. In fact, dust emissions are a threshold phenomenon, sporadic and spatially heterogeneous. Efforts have been deployed for more than 10 years to develop explicit dust emission models based on the physical processes at the interface between the atmosphere and the surface. Despite current modelling limits, the available emission models already allow to compute the dust flux produced for erodible arid surfaces satisfyingly provided the required input parameters (surface, soil and meteorological features) have been specifically determined. The development of accurate surface and soil databases and the use

© 2009 IOP Publishing Ltd

of the most pertinent meteorological datasets at the relevant scales are required to model mineral dust correctly.

2. Explicit and process-oriented dust emission models

2.1. Parameterizations of the erosion threshold

Dust emissions occur when the wind velocity (U), and thus the wind friction velocity (U^*), exceeds a certain value. This wind friction velocity value is called the threshold wind friction velocity (U^*_t). U^*_t controls both the occurrence of dust emissions and their intensity. Based on experimental data from [7] and [8], U^*_t can be considered as a function of the soil grain diameter. For large soil grains, U^*_t increases when the grain size increases (due to the gravity forces). For the smallest soil particles, U^*_t increases when the grain size decreases (mainly due to the inter-particle cohesive forces reinforcing grain bonds) [9]. These two effects lead to an optimum grain size around 80 μm for which U^*_t is the lowest (figure 1).

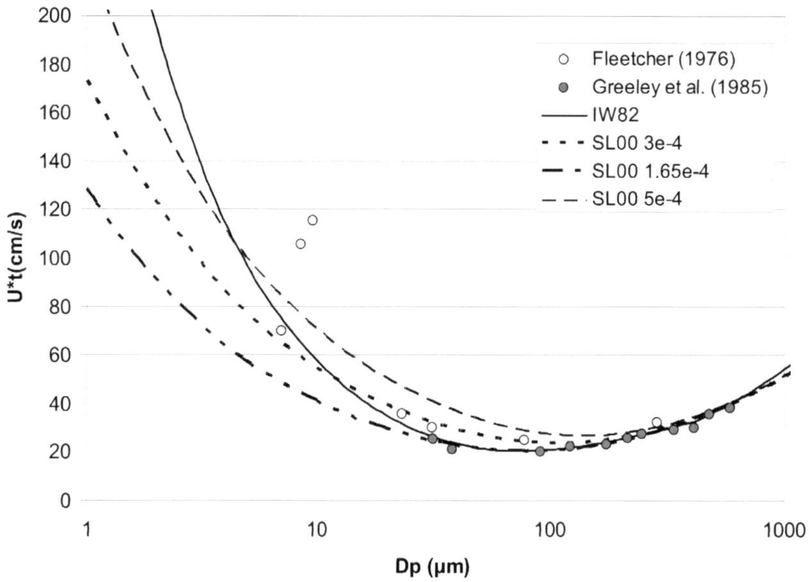

Figure 1: Threshold wind friction velocity (U^*_t) as a function of the soil aggregate diameter (D_p). Circles represent data from [10,11] and [12]. Simulations for several parameterizations: [9] (IW82), and [13] with $\gamma = 3 \times 10^{-4}$ kg.s^{-2} (SL00 3e-4), $\gamma = 1.65 \times 10^{-4}$ kg.s^{-2} (SL00 1.65e^{-4}), $\gamma = 5 \times 10^{-4}$ kg.s^{-2} (SL00 5e^{-4}).

Moreover, non erodible elements (pebbles, stones or vegetation), present on the desert surfaces, dissipate a part of the wind momentum that will not be available to initiate particle motion. This leads to a global decrease of the wind shear stress acting on the erodible surface and to an apparent increase of U^*_t. For example, [14] developed and validated a physical scheme describing the so-called drag partition (f_{eff}) between the roughness elements (characterized by the aerodynamic roughness length, Z_0) and the erodible surface (characterized by the smooth roughness length, z_{0s}). U^*_t can be then parameterized in "rough" or "smooth" situations which are encountered in arid and semi-arid areas:

$$U^*_t(D_p, Z_0, z_{0s}) = \frac{U^*_t(D_p)}{f_{eff}(Z_0, Z_{0s})} \quad \text{with} \quad f_{eff}(Z_0, z_{0s}) = 1 - \left(\left(\frac{Z_0}{z_{0s}}\right) \Big/ \left(0.35\left(\frac{10}{z_{0s}}\right)^{0.8}\right) \right) \tag{1}$$

The soil moisture (w) is another factor affecting U^*_t. The cohesive forces between the soil grains are reinforced by the soil water, and thus the erosion threshold is increased. A parameterization of this influence of the soil moisture on the erosion threshold was proposed by [15] (equation 2). The increase of the erosion threshold under wet conditions in reference to dry conditions is computed as a function of w and the residual soil moisture (maximum amount of adsorbed water, w', defined as a function of the soil clay content):

$$\text{For } w > w', \quad \frac{U^*_{tw}}{U^*_{td}} = \left[1 + 1.21\left(w - w'\right)^{0.68}\right]^{0.5} \quad \text{with } w' = 0.0014(clay\%)^2 + 0.17(clay\%) \tag{2}$$

2.2. Erosion horizontal flux (G)

Many theoretical and experimental works [7,16,17,18,19,20,21] showed that the horizontal flux (G), mostly composed of soil grains in movement of saltation or creeping, is proportional to the third power of U^*. Using a size-dependent expression of U^*_t, the *White*'s formulation also provides a size-dependent equation of G (equation 3) [22]. The amount of material mobilized by wind and its size-distribution can be then computed as a function of U^* (E is the fraction of erodible surface, $S_{rel}(D_p)$ is the relative surface covered by the particles of diameter D_p):

$$G = E\frac{\rho_a}{g}U^{*3}\sum_{D_p}\left(1 + \frac{U^*_t(D_p, Z_0, z_{0s})}{U^*}\right)\left(1 - \frac{U^*_t(D_p, Z_0, z_{0s})^2}{U^{*2}}\right)S_{rel}(D_p) \tag{3}$$

[22] showed that this size-dependent representation of G well-reproduces the total mass and the size distribution of the horizontal fluxes measured in a wind tunnel for different soil substrates and wind friction velocities by [23] or [19]. However, it must be noted that such a parameterization does not correctly reproduce the horizontal flux for supply-limited surfaces such as crusted soils [24].

2.3. Dust vertical flux (F)

The dust vertical flux *(F)* is composed of the finest particles setting in suspension from the saltation layer. These particles are then able to be transported over long distances. The dust production occurs during the "sandblasting" process, when the saltating grains impact on the surface and break the inter-particle bonds linking dust particles together or to the surface. Based on [17] coupled measurements of horizontal fluxes and vertical fluxes of dust particles with a diameter < 20 µm, [14] established an empirical relationship between the ratio F/G (i.e. the sandblasting efficiency, α) and the soil clay content. Using this parameterization the total mass of F can be estimated. Only sandblasting-process models allow to simulate both the mass and the size distribution of F explicitly.

[25] and [26] proposed a sandblasting-process model based on the existence of three typical dust particle populations that can be released from arid soils. These three modes are log-normally distributed and are characterized by specific binding energies corresponding to the thresholds of disruption of the soil aggregates. Dust of a given size is produced when the kinetic energy of the saltating soil particles exceeds the corresponding threshold. This implies that the dust size distribution varies as a function of the saltating particle size and their velocity [27,28].

In contrast to this energy-based model, the dust production can also be estimated as a function of the volume of soil removed by the saltating grains impacting on the surface [29]. An elaborated version of this model taking the saltation bombardment, the aerodynamic entrainment, and the aggregate disintegration into account, was developed by [30,31]. Large scale application of these sandblasting models is not easy due to the required input data (binding energy, soil plastic pressure) which cannot be easily determined from measurements at such a scale. This is currently one of the limitations to explicitly describe (i.e. based on physical processes) the size distribution of emitted mineral dust.

Major uncertainties in the modelling of dust emissions are the effects of the crusted soils and of specific soils such as the diatomite soil in the Bodélé Depression, the supply limitation of the soil, and the connection between the soil and the emitted dust size distribution and the dust composition. Moreover, the specific processes of dust emissions from the semi-arid vegetated and cultivated surfaces will need further investigations of the aerodynamic properties of the vegetation (porosity, flexibility, arrangement), of their temporal and spatial variability, and of the land use and agricultural practices. Additional data collection from these specific areas and new parameterization developments will be required.

3. Modelling dust emissions for regional and continental arid areas

The current explicit emission models allow to compute the dust flux emitted from erodible arid areas satisfyingly, provided input parameters have been specifically determined. For continental scale applications, the development and the use of accurate databases (surface, soil, and meteorology) are prerequisites to model mineral dust correctly.

3.1. Retrieval of relevant input data for the simulation of dust emissions

For the last decade, input databases have been developed alongside to the advancements of explicit dust emission models [e.g. 22,32,33,34,35,36,37,38]. We discuss here some of the most important soil, surface and meteorological input parameters for the modelling of dust emissions and some of the current methods used for their large scale establishments. It can already be noticed that a problem shared by all these methods is an under-constraint of the datasets due to the few available in-situ measurements.

In natural soil, various soil grain or aggregate sizes are present simultaneously. Since theses grains of different sizes have different threshold wind friction velocities, the in-situ soil size distribution has to be correctly described in emission models.

Soil maps generally classify soils according to the "textural triangle" defined by three size components: sand (2000 to 80 μm diameter), silt (80 to 4 μm diameter) and clay (<4 μm diameter) [39]. This classification is based on measurements performed by wet sedimentation techniques which break the soil aggregates (ultrasonic pre-treatment, dissolution). This leads to relatively high amounts of loose clay particles that generally form aggregates of larger size (>50-100 μm) and are then not encountered in the natural soils [40]. Moreover, [37] pointed out that for the north-east Asian deserts there is no direct relation between the soil texture and the soil grain size distribution. As a result, a soil texture classification cannot be directly used to characterize in-situ size distributions of erodible soils.

An alternative approach is to determine the soil size distribution using dry techniques that minimize, as much as possible, the breaking of the aggregates [40]. This approach was used by [39] for the characterization of the Saharan and Sahelian soils and applied to Chinese soil samples by [41]. In these works, the soil grain size is assumed to be log-normally distributed or to be the combination of 2 or 3 different log-normal distributions. However, the present data sets of such measurements remain limited and not homogeneous in terms of analytical methods. As mentioned by [40], there is a clear need for both standardized methods and sampling programs allowing for a correct mapping of the in situ size distribution of desert erodible soils.

The use of the drag partition scheme application for regional to global scale applications induces a similar problem to that encountered for the mapping of the soil grain size distributions: How can a map of aerodynamic roughness lengths (Z_0) be obtained over continental desert areas? A first method was developed for the Sahara desert by [22] and [32] based on a geomorphologic study of the surfaces. However, this method is very time consuming (with respect to the analysis of the data for continental areas) and its extension to other deserts depends on the number and quality of the available geomorphologic information.

A second method was to examine how satellite observations could provide a global mapping of Z_0. Following [42], who used the radar backscatter signal to retrieve Z_0 over selected desert targets, [35] investigated the possibility to retrieve Z_0 over arid areas using the surface products derived from the spaceborne POLDER-1 instrument (Polarization and Directionality of the Earth´s Reflectances). An empirical relationship between Z_0 and the protrusion coefficient (*PC*) derived from the POLDER-1 bi-directional reflectance distribution function in the visible range was determined. Maps of Z_0 (with a resolution of $^1/_{16}°\times^1/_{16}°$ to $¼°\times¼°$) were then established for the North African and north-east Asian deserts [35,38,43] (figure 2).

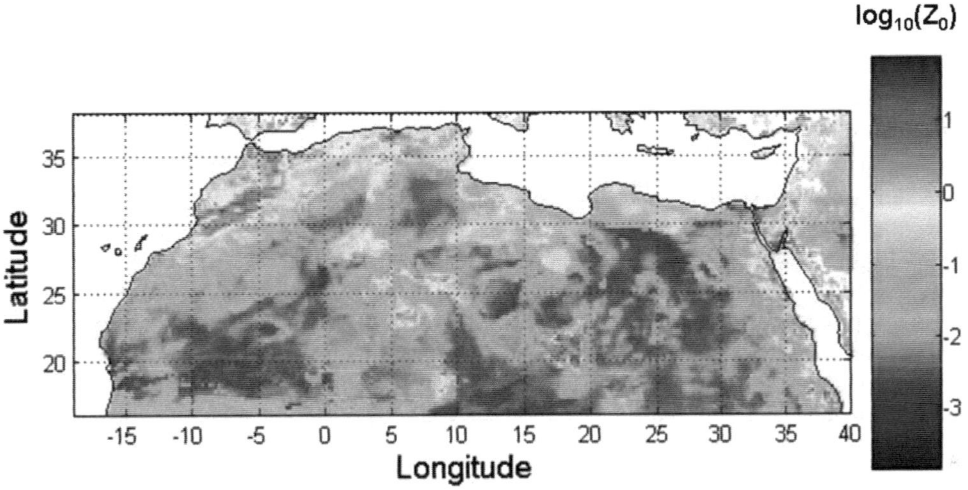

Figure 2: Map of $\log_{10}(Z_0)$ derived from POLDER-1 products and completed using a geomorphologic approach at the $¼°\times¼°$ spatial resolution (Z_0 is in cm, data available on: www.lisa.univ-paris12.fr).

Based on in-situ measurements of Z_0 in Tunisia, this approach was extended to the radar backscatter coefficient from high resolution images (Synthetic Aperture Radar, SAR, onboard the Earth Resource Satellite, ERS) [44]. This recent works provide an operational tool to derive Z_0 maps for local to regional applications. The variability of erodible desert surfaces in dust models will be better described using the physical parameterizations of the effects of Z_0 on the threshold wind friction velocities combined with these new datasets.

As already mentioned, dust emissions have a non linear dependence to wind speeds. Small errors on the surface wind speed values could lead to large errors on dust emissions [40]. Thus, the accuracy of the wind speed used in dust models is crucial. Most of the dust models use wind fields provided by meteorological centres (e.g. NCEP, ECMWF). Their spatial resolution is generally not higher than about 1°×1°. Other models, for instance those computing feedbacks between dust and dynamics, use wind fields computed on-line by an atmospheric general circulation model (with a lower resolution generally of 2°×2° up to 5°×5°). In certain locations as the Bodélé Depression, low-resolved wind speeds tend to be underestimated [45], and they can not reproduce specific meteorological processes forcing dust emissions (e.g. low-level jet, density current, squall line). Additional parameterizations, mainly based on wind probability density functions, were introduced to account for the subgrid wind fluctuations [46,47,48]. If such an approach is valuable for climatic simulations, the use of wind fields derived from well-resolved regional meteorological models (e.g. MM5, RAMS, Lokal Modell) should be favoured for simulations of specific dust events [40].

3.2. Regional to continental simulations of dust emissions

The explicit dust emission model developed by [14] and coupled with relevant input datasets of the surface characteristics allows to simulate the dust emission occurrences and their intensity over continental desert areas with a good confidence level [37,38]. Both the emission occurrences and the intensity have to be modelled correctly, the radiative and biogeochemical impacts of sporadic but intense dust pulse events on the environment being different from the dust background impact due to frequent but less intense events.

Using this model configuration, dust emissions from the Sahara over a 6-yr simulated period correspond to an average value of 670 Tg ± 60 Tg per year with pronounced seasonal cycles [38]. These Saharan dust emissions are 2 or 3 times higher than the emissions simulated under the same conditions for the north-east Asian deserts (240 Tg ± 130 Tg) [37]. Using the modelling method developed in these two studies, dust emitted from large source areas over North African and north-east Asian deserts but also from small and very active areas ("hot spots") are simulated. In addition, the sub-daily temporal resolution of the simulations respects the sporadic nature of dust emissions and allows to reproduce non-frequent and unusually intense dust events which can partly control the interannual variability of the dust emissions.

Studies have also been done to evaluate the performance of regional dust model systems (e.g. [49,50,51]). Occurrences of significant dust emissions (i.e. dust flux > 10^{-10} $g.cm^{-2}.s^{-1}$) simulated with the LM-MUSCAT regional model for two periods of the Saharan Mineral Dust Experiment (SAMUM) with a spatial resolution of 28 km×28 km and a hourly temporal resolution are presented in figures 3 a and c. The spatio-temporal variability of the simulated emissions is in agreement with the dust emission observed from Meteosat Second Generation (MSG) SEVIRI IR difference images and using a back-tracking approach developed by [52] for the same two periods (figures 3 b and d respectively). The emissions observed with MSG (only the first ones when there is no cloud and no dust stagnation) correspond to a minimum frequency of occurrence. While regional models generally provide good results in reproducing the spatio-temporal dust variability for cases where dust emission is caused by large-scale dynamics, dust emissions in connection with moist convective events can, for instance, be misrepresented [51]. Currently, one of the main questions and uncertainties concerns the regional evaluation and forecast of emissions from semi-arid sources, as for example in the Sahel. The improvement of dust emission modelling and forecasting are necessary to determine and manage the risks and nuisances for the exposed populations living in these areas correctly.

The validation of the simulated dust emissions is also a complex issue due to the lack of quantitative measurements directly over desert surfaces. The appropriate strategy to test the simulations should be to use all available proxy of mineral dust content close to the source areas and during transport (horizontal visibility measurements (e.g. [37,53,54]); aerosol optical thickness from AERONET, on http://aeronet.gsfc.nasa.gov/; vertical profiles from EARLINET network, on http://www.earlinet.org/). The validation should also be based on the use of the satellite products over continental (aerosol index from OMI TOMS and Meteosat/IRT; dust emission observation from MSG) and oceanic surfaces (optical depth from SeaWIFS, Meteosat/VIS, PARASOL; vertical distribution from CALIPSO). As already mentioned, [52] made use of 3 thermal IR wavelength channels of the MSG SEVIRI instrument to detect atmospheric dust over land. The high spatio-temporal resolution renders then the detection of location of individual dust emission events possible (figures 3 b and d). Successful field campaigns which have studied mineral dust (e.g. African Monsoon Multidisciplinary Analyses, AMMA; Saharan Mineral Dust Experiment, SAMUM; Bodélé Dust Experiment, BoDEx) will be useful to better understand the mineral dust properties and to validate dust simulations.

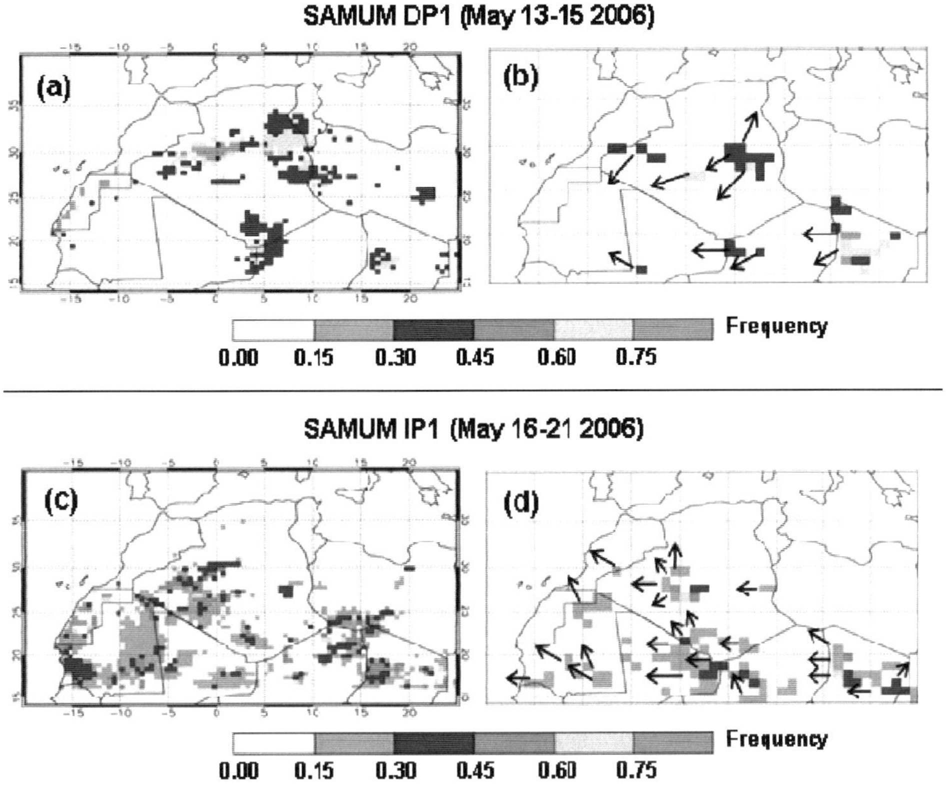

Figure 3: Occurrence of significant dust emissions (i.e. dust flux > 10^{-10}g.cm^{-2}.s^{-1}) simulated with the LM-MUSCAT regional model (a), and occurrence of dust emissions observed with Meteosat Second Generation (b) during SAMUM Dust Phase 1 (DP1). The same informations are reported for the SAMUM Intermediate Phase 1 (c and d respectively). For the MSG observations, only the first pixel (1°×1°) where dust emission is observed is taken into account and compiled in the maps (when there is no cloud and no dust stagnation). The arrows show the main transport directions of the dust plumes under which emitted and transported dust cannot be differentiated.

4. Conclusion

During the last decade, dust emission models based on a physical basis and allowing for a better consideration of the heterogeneity of the surface features were developed. This is an important step ahead since the spatial variability of dust emissions is partly controlled by differences of the erodible surfaces in terms of soil grain size distributions, texture, and surface roughness. Alongside, new techniques have been developed (as ones derived from satellite observations) to provide relevant surface data required by the emission models. Maps of Z_0 have been developed for the Saharan and Asian deserts and a global map for all the arid desert areas should be available soon. Though new data on the soil grain size distributions of desert areas are required, the available datasets have significantly grown. Sandblasting models have been conceptually developed and should be operational soon. This represents a progress since these models have the capability to simulate the dust flux and the dust size distribution explicitly. Accurate size-resolved dust fluxes are key parameters to simulate the long-range transport and the impacts of dust correctly. However, as a consequence of the progresses in dust production parameterizations and in the surface databases used in dust models, the uncertainty due to the lacking accuracy of the surface wind speed becomes more and more evident. A large part of the future advances in dust modelling, especially to establish a forecasting system, will thus be focus on the provision of better resolved and more accurate wind fields for dust emission models. The new developments in satellite dust observations and the recent field campaigns will offer unique frames to better constrain dust emission simulations.

References

[1] Sokolik, I.N. and O.B. Toon, 1999 Incorporation of the mineralogical composition into models of the radiative properties of mineral aerosol from UV to IR wavelengths, *J. Geophys. Res.*, **104**, D8, 9423-9444.

[2] Loÿe-Pilot, M.D., J.M. Martin and J. Morelli, 1986 Influence of Saharan dust on the rain acidity and atmospheric input to the Mediterranean, *Nature*, **321**, 427-428

[3] Bauer, S.E., Y. Balkanski, M. Schulz, D.A. Hauglustaine and F. Dentener, 2004 Global modeling of heterogeneous chemistry on mineral aerosol surfaces: The influence on tropospheric ozone chemistry and comparison to observations, *J. Geophys. Res.*, **109**, D02304, doi: 10.1029/2003JD003868.

[4] Bergametti, G., E. Remoudaki, R. Losno, E. Steiner, B. Chatenet and P. Buat-Ménard, 1992 Sources, transport and deposition of atmospheric phosphorus over the northwestern Mediterranean, *J. Atmos. Chem.*, **14**, 501-513.

[5] Swap, R., M. Garstang, S. Greco, R. Talbot and J.Y. Gac, 1992 Sahara dust in the Amazon basin, *Tellus B*, **44**, 133-149.

[6] Jickells T.D., Z.S. An, K.K. Andersen, A.R. Baker, G. Bergametti, N. Brooks, J.J. Cao, P.W. Boyd, R.A. Duce, K.A. Hunter, H. Kawahata, N. Kubilay, J. La Roche, P.S. Liss, N. Mahowald, J.M. Prospero, A.J. Ridgwell, I. Tegen and R. Torres, 2005 Global iron connections: Between desert dust, ocean biogeochemistry and climate. *Science,* **308,** 5708, 67-71.

[7] Bagnold, R.A., 1941 The physics of blown sand and desert dunes, Methuen, New York, 265 pp.

[8] Chepil, W.S., 1945 Dynamics of wind erosion, *Soil Sci*, **60**, 305-320.

[9] Iversen, J.D. and B.R. White, 1982 Saltation threshold on Earth, Mars and Venus, *Sedimentology*, **29**, 111-119.

[10] Fletcher, B., 1976 The erosion of dust by an airflow, *J. Phys. D Appl. Phys.*, **9**, 17, 913-924.

[11] Fletcher, B., 1976 The incipient motion of granular materials, *J. Phys. D Appl. Phys.,* **9**, 17, 2471-2478.

[12] Greeley, R. and J.D Iversen, 1985 Wind as a geological process on Earth, Mars, Venus and Titan, Cambridge University Press, New-York, 333 pp.

[13] Shao., Y. and H. Lu, 2000 A simplified expression for threshold friction velocity, *J. Geophys. Res.,* **105**, 22,437-22,443.

[14] Marticorena, B. and G. Bergametti, 1995 Modeling the atmospheric dust cycle: 1-Design of a soil derived dust production scheme, *J. Geophys. Res.,* **100**, 16415-16430

[15] Fécan, F., B. Marticorena and G. Bergametti, 1999 Parameterization of the increase of the aeolian erosion threshold wind friction due to soil moisture for semi arid areas, *Ann. Geophys.,* **17**, 149-157.

[16] White, B.R., 1979 Soil transport by winds on Mars, *J. Geophys. Res.,* **84**, 4643-4651.

[17] Gillette, D.A., 1979 Environmental factors affecting dust emission by wind erosion in Saharan Dust, C. Morales (Ed.), John Wiley, New York, 71-94.

[18] Gillette, D.A. and P.H. Stockton, 1989 The effect of nonerodible particles on wind erosion of erodible surfaces, *J. Geophys. Res.,* **94***,* 12,885-12,893.

[19] Sörensen, M., 1985 Estimation of some aeolian saltation transport parameters from transport rate profiles, in Proceedings of the International Workshop on the Physics of Blown Sand, O.E. Barndorff- Nielsen, J.T. Möller, K. Römer Rasmussen, B.B. Willets (Eds.), University of Aarhus, Aarhus, Denmark, 141-190.

[20] Leys, J.F. and M.R. Raupach, 1991 Soil flux measurements with a portable wind erosion tunnel, *Aust. J. Soil Res.,* **29**, 533-552.

[21] Shao, Y., M.R. Raupach and P.A. Findlater, 1993 Effect of saltation bombardment on the entrainment of dust by wind, *J. Geophys. Res.*, **98**, 12,719-12,726.

[22] Marticorena, B., G. Bergametti, B. Aumont, Y. Callot, C. N'Doumé and M. Legrand, 1997 Modeling the atmospheric dust cycle: 2-Simulation of Saharan sources, *J. Geophys. Res.,* **102**, 4387-4404.

[23] Williams, G., 1964 Some aspects of the aeolian saltation load, *Sedimentology,* **3**, 253-256.

[24] Lopez, M.V., 1998 Wind erosion in agricultural soil: an example of limited supply of particles available for erosion, *Catena*, **33**, 17-28.

[25] Alfaro, S., A. Gaudichet, L. Gomes and M. Maillé, 1997 Modeling the size distribution of a soil aerosol produced by sandblasting, *J. Geophys. Res.,* **102**, 11,239-11,249.

[26] Alfaro, S., A. Gaudichet, L. Gomes and M. Maillé, 1998 Mineral aerosol production by wind erosion: aerosol particles size and binding energies, *Geophys. Res. Lett.,* **25**, 991-994.

[27] Alfaro, S. and L. Gomes, 2001 Modeling mineral aerosol production by wind erosion: emission intensities and aerosol size distributions in source areas, *J. Geophys. Res.,* **106**, 18,075-18,084.

[28] Alfaro, S., J.L. Rajot and W.G. Nickling, 2004 Estimation of PM20 emissions by wind erosion: main sources of uncertainties, *Geomorphology,* **59**, 63-74.

[29] Lu, H. and Y. Shao, 1999 A new model for dust emission by saltation bombardment, *J. Geophys. Res.,* **104**, 16,827–16,842.

[30] Shao, Y., 2001 A model for mineral dust emission, *J. Geophys. Res.,* **106**, 20,239–20,254.

[31] Shao, Y, 2004 Simplification of a dust emission scheme and comparison with data, *J. Geophys. Res.,* **109**, doi:10.1029/2003JD004372.

[32] Callot, Y., B. Marticorena and G. Bergametti, 2000 Geomorphologic approach for modelling the surface features of arid environments in a model of dust emissions: application to the Sahara desert, *Geodin. Acta,* **13**, 245-270.

[33] Tegen, I., S.P. Harrison, K. Kohfeld, I.C. Prentice, M. Coe and M. Heimann, 2002 Impact of vegetation and preferential source areas on global dust aerosol: Results from a model study, *J. Geophys. Res.,* **107**, doi: 10.1029/2001JD000963.

[34] Zender, C.S., D. Newman and O. Torres, 2003 Spatial heterogeneity in aeolian erodibility: Uniform, topographic, geomorphic, and hydrologic hypotheses, *J. Geophys. Res.,* **108**, D17, doi:10.1029/2002JD003039.

[35] Marticorena, B., P. Chazette, G. Bergametti, F. Dulac and M. Legrand, 2004 Mapping the aerodynamic roughness length of desert surfaces from the POLDER/ADEOS bi-directional reflectance product, *Int. J. Remote Sens.,* **25**, 603-626.

[36] Prigent C., I. Tegen, P. Aires, B. Marticorena and M. Zribi, 2005 Estimation of the aerodynamic roughness length in arid and semi-arid regions over the globe with the ERS scatterometer, *J. Geophys. Res.,* **110**, D09205, doi: 10.1029/2004JD005370.

[37] Laurent, B., B. Marticorena, G. Bergametti and F. Mei, 2006 Modeling mineral dust emissions from Chinese and Mongolian deserts, *Global Planet. Change,* **52**, 1-4, 121-141.

[38] Laurent, B., B. Marticorena, G. Bergametti, J.F. Léon and N.M. Mahowald, 2008 Modeling mineral dust emissions from the Sahara desert using new surface and soil developments, *J. Geophys. Res.,* doi:10.1029/2007JD009484.

[39] Chatenet, B., B. Marticorena, L. Gomes and G. Bergametti, 1996 Assessing the microped size distributions of desert soils erodible by wind, *Sedimentology,* **43**: 901-911.

[40] Bergametti, G., B. Marticorena and B. Laurent, 2007 Key processes for dust emissions and their modeling, in Regional climate variability and its impacts in the Mediterranean area, A. Mellouki and A.R. Ravishankara (Eds.), Nato Science Series: IV: Earth and Environmental Sciences, 79, 326 pp.

[41] Mei, F., X. Zhang, H. Lu, Z. Shen and Y. Wang, 2004 Characterization of MASDs of surface soils in north China and its influence on estimating dust emission, *Chin. Sci. Bull.*, **49**, 2169–2176.

[42] Greeley, R., D.G. Blumberg, J.F. McHone, A. Dobrovolski, J. Iversen, N. Lancaster, K.R. Rasmussen, S. Wall and B. White, 1997 Applications of spaceborne radar laboratory data to the study of aeolian processes, *J. Geophys. Res.*, **102**, 10,971-10,983.

[43] Laurent, B., B. Marticorena, G. Bergametti, P. Chazette, F. Maignan and C. Schmechtig, 2005 Simulation of the mineral dust emission frequencies from desert areas of China and Mongolia using an aerodynamic roughness length map derived from the POLDER/ADEOS 1 surface products, *J. Geophys. Res.*, **110**, D18S04, doi: 10.1029/2004JD005013.

[44] Marticorena, B., M. Kardous, G. Bergametti, Y. Callot, P. Chazette, H. Khatteli, S. Le Hégarat-Mascle, M. Maillé, J.L. Rajot, D. Vidal-Madjar and M. Zribi, 2006 Aeolian geometric and aerodynamic surface roughness in arid and semi-arid areas and their relation with radar backscatter coefficient, *J Geophys. Res.*, **111**, doi: 10.1029 /2006JF000462.

[45] Bouet, C., G. Cautenet, R. Washington, M.C. Todd, B. Laurent, B. Marticorena and G. Bergametti, 2007 Mesoscale modeling of aeolian dust emission during the BoDEx 2005 experiment, *Geophys. Res. Lett.*, **34**, L07812, doi:10.1029/2006GL029184.

[46] Westphal, D.L., O.B. Toon and T.N. Carlson, 1988 A case study of mobilization and transport of Saharan dust, *J. Atmos. Sci.*, **45**, 2145– 2175.

[47] Gillette, D.A. and R. Passi, 1988 Modeling dust emission caused by wind erosion, *J. Geophys. Res.*, **93**, 14,233-14,242.

[48] Cakmur, R.V., R.L. Miller and O. Torres, 2004 Incorporating the effect of small-scale circulations upon dust emission in an atmospheric general circulation model, *J. Geophys. Res.*, **109**, doi: 10.1029/2003JD004067.

[49] Shao, Y., Y. Yang, J. Wang, Z. Song, L.M. Leslie, C. Dong, Z. Zhang, Z. Lin, Y. Kanai, S. Yabuki and Y. Chun, 2003 Northeast Asian dust storms: Real-time numerical prediction and validation, *J. Geophys. Res.*, **108**, 4691, doi:10.1029/2003JD003667.

[50] Pérez, C., S. Nickovic, J.M. Baldasano, M. Sicard, F. Rocadenbosch and V.E. Cachorro, 2006 A long Saharan dust event over the western Mediterranean: Lidar, Sun photometer observations, and regional dust modeling, *J. Geophys. Res.*, **111**, D15214, doi:10.1029/2005JD006579.

[51] Heinold, B., I. Tegen, M. Esselborn, K. Kandler, P. Knippertz, D. Müller, A. Schladitz, M. Tesche, B. Weinzierl, A. Ansmann, D. Althausen, B. Laurent, A. Massling, T. Mueller, A. Petzold, K. Schepanski and A. Wiedensohler, 2008 Regional Saharan dust modelling during the SAMUM 2006 campaign, *Tellus B*, doi: 10.1111/j.1660-0889.2008.00387.

[52] Schepanski, K., I. Tegen, B. Laurent, B. Heinold and A. Macke, 2007 A new Saharan dust source activation frequency map derived from MSG-SEVIRI IR-channels, *Geophys. Res. Lett.*, **34**, L18803, doi:10.1029/2007GL030168.

[53] Middleton, N.J., 1989 Climatic controls on the frequency, magnitude and distribution of dust storms: examples from India/Pakistan, Mauritania and Mongolia, M. Leinen and M. Sarnthein (Eds.), Paleoclimatology and paleometeorology: Modern and past patterns of global atmospheric transport, Kluwer Academic Publ., Dordrecht, 97–132.

[54] Mahowald, N.M., J.A. Ballantine, J. Feddema and N. Ramankutty, 2007 Global trends in visibility: Implications for dust sources, *Atmos. Chem. Phys.*, **7**, 3309-3339.

IOP Conf. Series: Earth and Environmental Science 7 (2009) 012007 doi:10.1088/1755-1307/7/1/012007

Operational Aerosol and Dust Storm Forecasting

D L Westphal, C A Curtis, M Liu, and A L Walker

Naval Research Laboratory. Monterey, USA

Email: westphal@nrlmry.navy.mil

Abstract. The U. S. Navy now conducts operational forecasting of aerosols and dust storms on global and regional scales. The Navy Aerosol Analysis and Prediction System (NAAPS) is run four times per day and produces 6-day forecasts of sulfate, smoke, dust and sea salt aerosol concentrations and visibility for the entire globe. The Coupled Ocean Atmosphere Mesoscale Prediction System (COAMPS®) is run twice daily for Southwest Asia and produces 3-day forecasts of dust, smoke, and visibility. The graphical output from these models is available on the Internet (www.nrlmry.navy.mil/aerosol/). The aerosol optical properties are calculated for each specie for each forecast output time and used for sea surface temperature (SST) retrieval corrections, regional electro-optical (EO) propagation assessments, and the development of satellite algorithms. NAAPS daily aerosol optical depth (AOD) values are compared with the Advanced Very High Resolution Radiometer (AVHRR) and Moderate Resolution Imaging Spectroradiometer (MODIS) AOD values. Visibility forecasts are compared quantitatively with surface synoptic reports.

1. Introduction

A number of naturally occurring and anthropogenic aerosols affect radiative transfer and visibility in all parts of the world. In the past, forecasters relied on a combination of diagnostic tools for estimating localized aerosol effects on sensors and systems. No operational predictive capability existed. Coincidentally, the academic community is also interested in the global distributions of natural and anthropogenic aerosols because of their potential climatic and health implications.

The Naval Research Laboratory in Monterey (NRL/MRY) is involved in a wide range of aerosol research and development activities designed to meet the Navy's objective of aerosol simulation and forecasting for use in predicting visibility. NRL has developed, tested, and transitioned to operational use at the Fleet Numerical Meteorology and Oceanography Center (FNMOC) a new capability for forecasting the concentration of atmospheric particulate matter and the subsequent effects on visibility for the entire globe. The forecasts are used by a wide range of users, including weather forecasters, mission planners, operators, and scientists.

2. Global Forecasting

The Navy Aerosol Analysis and Prediction System (NAAPS) global aerosol model was developed at the Naval Research Laboratory in Monterey, USA. NAAPS is a global three-dimensional aerosol and air pollution model, based primarily on the Danish Eulerian Hemispheric Model (DEHM; [1]). NAAPS is an offline model that utilizes the meteorological analysis and forecast fields from the 0.5 degree Navy's Operational Global Analysis and Prediction System (NOGAPS; [2]; Hogan and Brody, 1993). Real-time forecasting began in 1999 with twice-daily forecasts for sulfate and mineral dust

© 2009 IOP Publishing Ltd

with an extensive display of model products on the web, including side-by-side comparison with satellite, surface observations, and Aerosol Robotic Network (AERONET) data. A biomass smoke component was added in 2001, utilizing the real-time smoke sources provided by the Fire Locating and Modeling of Burning Emissions (FLAMBE; [3]). NAAPS became fully operational at the Fleet Numerical Meteorology and Oceanography Center on October 2004 becoming the world's first operational global aerosol model. A sea salt component was added in 2006 [4]. Presently, NAAPS produces 6-day forecasts of SO2, sulfate, dust, biomass burning smoke and sea salt mass concentration with 1x1 degree resolution at 25 levels. A data assimilation system has been developed and is now being transitioned to operations [5]. The forecasts are used by a wide range of users, including weather forecasters, mission planners, operators, and scientists. NAAPS forecasts are used to correct satellite retrievals of sea surface temperature (SST) by European Organisation for the Exploitation of Meteorological Satellites (EUMETSAT) and the Naval Oceanographic Office (NAVO). The model output is further processed and converted into the fundamental optical properties required to calculate electro-optical propagation.

FNMOC produces two six-day NAAPS forecasts each day. The forecasts of aerosol concentration are distributed via the internet for use by forecasters, air quality agencies, and scientists. The forecasts are used to correct satellite retrievals of SST by NAVO, thus improving hurricane forecasts. The model output is further processed and converted into the fundamental optical properties required to calculate EO propagation. These properties are used to populate the Tactical Environmental Data Server (TEDS) and by the Target Acquisition Weapons Software (TAWS) to calculate slant-path visibility. NAAPS has been used and referenced in over 50 scientific papers and meeting presentations.

3. Regional Forecasting

A 3-dimensional mesoscale aerosol model has been developed at NRL/MRY [6]. It is implemented in the US Navy's regional numerical weather forecast system COAMPS [7,8] as an on-line module fully coupled at each time step and grid point and on each nested grid. It simulates the evolution of the spatial and size distributions of mineral dust particles and passive volcano ash. A 5th-order positive-definite flux-form advection scheme is used in both the horizontal and the vertical along with a semi-implicit mixing scheme of turbulent kinetic energy closure to achieve conservative mass and free dispersion. The aerosol physical processes of sedimentation, dry deposition, and wet removal are calculated using the dynamics and cloud fields of COAMPS. A high-resolution dust source database has been developed based on satellite remote sensing, land use and cover information [9] to match the resolution of COAMPS. Biomass smoke and sea salt aerosols are being permanently added to COAMPS. The source emissions of volcano ash and biomass smoke are automatically determined from real-time satellite data [3]. The COAMPS aerosol model has successfully simulated Asian dust storms with 10 particle bins ranging from 0.05 to 35 um in diameters. It has also successfully predicted volcanic ash plumes and biomass smoke plumes.

Real-time dust forecasting began during Operation Iraqi Freedom (OIF) in March and April 2003. Since that time COAMPS has been operationally producing 3-day forecasts of dust aerosol for Southwest Asia, particularly in the Iraq region where dust storms frequently occur [10]. The operational domain has three nested grid meshes of 6, 18, and 54 km with the innermost domain centered in the area of interest. A two-way nest interaction is performed to allow transport of dust plumes across grid mesh boundaries. The dust forecasts are made twice a day starting at 00Z and 12Z with a 6-hour update cycle of meteorological data assimilation, while dust mass is continuously simulated throughout time. COAMPS generates dust mass concentration, mass load, extinction coefficient and optical depth, as well as visibility (Fig. 1.). The daily products are posted on the NRL/MRY aerosol website www.nrlmry.navy.mil/aerosol/.

4. Dust Source Specification

In many dust models, the dust source areas for the globe have been derived from an analysis of satellite data at one-degree latitude/longitude resolution. Due to the new capability of high-resolution weather modeling with COAMPS, we have developed a high-resolution source term for Southwest Asia and East Asia [9]. A Geographic Information Systems (GIS) is used to derive the new source map from MODIS satellite imagery as well as written reports, surface observations and other relevant information. This approach is invaluable for the high-resolution dust storm modeling using COAMPS. The approach will soon be extended to Africa.

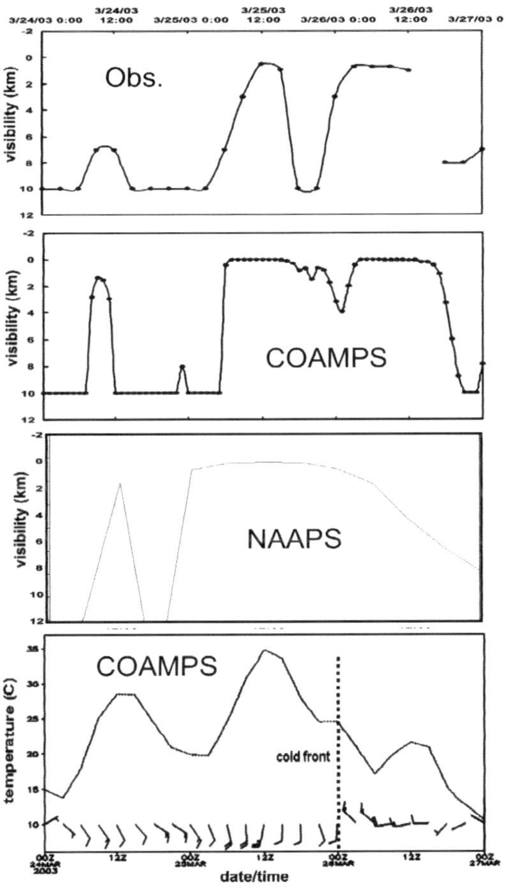

Figure 1. Time series of observed visibility, COAMPS and NAAPS 0–72-h forecasts of dust visibility and COAMPS 0–72-h forecasts of surface temperature and wind barbs (a full tick = 10 m/s) at Arar weather station from 24 to 27 Mar 2003. Note the inverted visibility (y) axis.

5. Significance

With the global and regional models, we are now able to predict the concentration of the dominant visibility reducing aerosol species up to six days in advance anywhere on the globe. Previous to NAAPS, a simple climatology or 'rules of thumb' were all that was available for forecasters. NAAPS is particularly useful for forecasts of dust transport from the large deserts of the world to the maritime environment: Sea of Japan, China Sea, Mediterranean Sea, and the Tropical Atlantic Ocean [11]. NAAPS also accurately predicts the fate of large-scale smoke plumes originating from boreal and tropical forests and the savannah [12,13].

6. Validation

The model performance is evaluated using surface weather reports, visibility observations, and enhanced satellite retrievals. The testing of NAAPS involves comparison of aerosol concentrations and optical effects against the available observations. These include conventional surface synoptic reports of visibility, and current and past weather, global networks of sun photometers, and lidar. The observations also come from intensive field studies, some of which have been led by NRL/MRY.

The comparison of the COAMPS dust forecasts with observations for the severe dust storms of OIF shows that COAMPS predicted the arrival and retreat of the major dust events within 2 h ([10]; Fig. 1.) In most cases, COAMPS predicted the intensity (reduction in visibility) of storms with an error of less than 1 km. The forecasts of the spatial distribution of dust fronts and dust plumes were consistent with those seen in the satellite images and the corresponding cold front observations.

A statistical analysis of dust-related visibility for the OIF period reveals that COAMPS generates higher bias, root mean square error (RMSE), and relative errors at the stations having high frequencies of dust storms and near the source areas. [10] also validates the COAMPS forecast using prediction rates taken from quantitative precipitation forecasting and prediction rates familiar to weather forecasters. These are:

- dust storm prediction rate: ratio of the number of correctly predicted dust incidents to the number of observed dust incidents,
- dust storm false alarm rate: ratio of the number of falsely predicted dust incidents to the number of observed clear-sky incidents,
- dust storm threat score: (number of predicted dust incidents)/(predicted dust + missed dust + false alarm dust incidents),
- total prediction rate: (number of correctly predicted dust incidents + correctly predicted clear-sky incidents)/(total observations).

The calculation of forecast accuracy (Fig. 2) shows that COAMPS achieved a probability of dust detection of 50%–90% and a threat score of 0.3–0.55 at the stations with frequent dust storms. Overall, the model predicted more than 85% of the observed dust and non-dust weather events at the stations used in the verification for the OIF period.

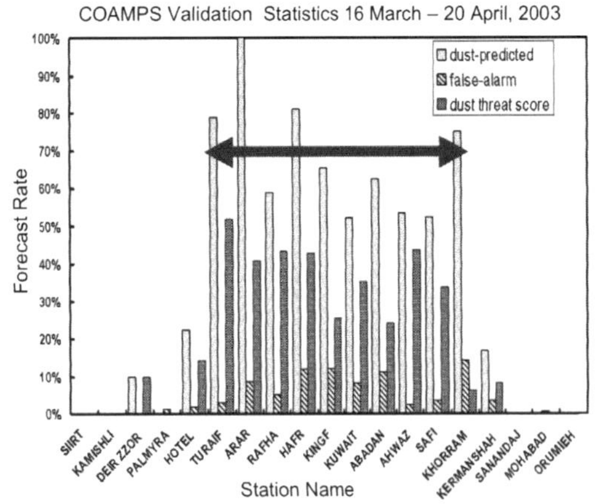

Figure 2. Forecast statistics for COAMPS predictions of dust storms (defined as conditions of horizontal visibility less than 3.5 km. Shown are the dust storm prediction rate, false alarm rate, and threat score for forecasts at 19 stations surrounding Iraq from 16 Mar to 20 Apr 2003. Double-ended arrow notes those stations typically downwind of Iraq sources of dust.

7. Conclusions

Our conclusions are presented in the form of our recommendations made relative to the issues posed by the organizers:

- *"Identify, present and suggest future real-time observations for forecast verification and dust surveillance: satellite, ground-based remote sensing (passive and active) and in-situ monitoring."* A practical reality facing modelers is that few people live in or near the dust sources and thus there are few data for validation. Contrast this to modeling studies of anthropogenic aerosols, which by definition occur and are observed in heavily populated areas. This situation poses a problem for model validation. Regardless, we need to do as much near-source data validation as is possible and fix problems with source functions, PBL development, and dynamics in the source areas before validating the models far downwind of the sources. The highest priority study might be the validation of the vertical distribution of dust in the source regions, which is now possible using CALIPSO data.

Studies that compare model forecasts with wet deposition data in Florida or vertical profiles over Northern Europe, as were presented at the WMO SDSWG meeting do educate the user to the skill of the forecasts. However, if the model fails we don't have the observations to understand why the model failed since many atmospheric and microphysical processes occur between the wadis of Central Africa and N. Europe and these are all under sampled.

- *"Discuss and identify user needs: health, air quality, air transport operations, ocean, and others."* We expect that the health effects of Saharan dust storms in Europe pales in comparison to that in Africa itself, yet there is little information and few suggestions of how model forecasts can be used to ameliorate these effects. The African health and meteorology agencies, and WMO health agencies need to better explain their needs. Concurrently, the meteorology community needs to better explain their forecast capabilities. This dialogue was initiated at the WMO SDSWG meeting but has a long way to go. The same recommendation applies for dust storms in East Asia. The problems of those living in Western China are more severe than for those living in Japan and Korea yet those in Western China have not explained their needs.

- *"Identify and discuss dust research issues relevant for operational forecast applications."* Those living near the sources can contribute to dust storm forecasting by providing real-time information regarding the time-dependence of dust sources. Static dust source databases are unsuitable for capturing the seasonal, annual, and interdecadal variations due to agriculture, flooding, drought, and climate change. Local agricultural and meteorological agencies could make a significant contribution in this area. Remote sensing approaches have shown some skill but need ground truth validation and need temporal updating.

Acknowledgments

The authors would like to thank the meeting organizers and the WMO for an excellent meeting and support for travel to the meeting. The authors gratefully acknowledge support for research from the Office of Naval Research and the Naval Research Laboratory through programs PE-061153N and PE-0602435N. FNMOC is very much appreciated for providing the computation resources to this work.

References

[1] Christensen J. H., 1997 The Danish Eulerian hemispheric model - A three-dimensional air pollution model used for the Arctic, *Atmos. Environ.*, **31** (24), 4169-4191.

[2] Hogan, T. F., and T. E. Rosmond, 1991 The description of the Navy Operational Global Atmospheric Prediction Systems spectral forecast model, *Mon. Wea. Rev.*, **119**(8), 1786-1815.

[3] Reid, J. S., E. M. Prins, D.L. Westphal, et al., 2004 Real-time monitoring of South American smoke particle emissions and transport using a coupled remote sensing/box-model approach, *Geophys. Res. Lett.*, **31**, L06107, doi:10.1029/2003GL018845.

[4] Witek, M. L., P. J. Flatau, P. K. Quinn, and D. L. Westphal, 2007 Global sea-salt modeling: Results and validation against multicampaign shipboard measurements, *J. Geophys. Res.*, **112**, D08215, doi:10.1029/2006JD007779.

[5] Zhang, J., J. S. Reid, D. L. Westphal, N. Baker, and E. J. Hyer, 2008 A system for operational aerosol optical depth data assimilation over global oceans, *J. Geophys. Res.*, **113**, doi:10.1029/2007JD009065.

[6] Liu, M., D. L. Westphal, S. Wang, A. Shimizu, N. Sugimoto, J. Zhou and Y. Chen, 2003 A high-resolution numerical study of Asian dust storms of April 2001. *J. Geophys. Res.*, Vol. **108**, (D23) 8653, doi:10.1029/2002JD003178.

[7] Hodur, R. M., 1997 The Naval Research Laboratory's Coupled Ocean/Atmosphere Mesoscale Prediction System (COAMPS), *Mon. Wea. Rev.*, **125**, 1414-1430.

[8] Chen, S., J. Cummings, J. Doyle, R.H. Hodur, T. Holt, C. Liou, M. Liu, A. Mirin, J. Ridout, J.M. Schmidt, G. Sugiyama, and W. T. Thompson, 2003 COAMPS 3.0 Model Description-General Theory and Equations. Naval Research Lab. NRL Tech Note NRL/PU/7500-0-3-448, May, 2003. 143 pp.

[9] Walker, A. L., M. Liu, S. D. Miller, K. A. Richardson, and D. L. Westphal, 2009 A High Resolution Dust Source Database for Southwest Asia, submitted to *J. Geophys. Res.*

[10] Liu, M., D. L. Westphal, A. L. Walker, T. R. Holt, K. A. Richardson and S. D. Miller, 2007 COAMPS real-time dust storm forecasting during Operation Iraqi Freedom. *J. Weather and Forecasting*, **22**, 192-206.

[11] Park, C. B., N. Sugimoto, I. Matsui, A. Shmizu, B. Tatarov, A. Kamei, C.H. Lee, I. Uno, T. Takemura, and D. L. Westphal, 2005 Long-Range Transport of Saharan Dust to East Asia Observed with Lidars, *SOLA*, 1.

[12] Jaffe, D., I. Bertschi, L. Jaeglé, P. Novelli, J. S. Reid, H. Tanimoto, R. Vingarzan, D. L. Westphal, 2004 Long-range transport of Siberian biomass burning emissions and impact on surface ozone in western North America, *Geophys. Res. Lett.*, **31**, L16106, doi:10.1029/2004GL020093.

[13] Honrath, R. E., R. C. Owen, M. Val Marti´n, J. S. Reid, K. Lapina, P. Fialho, M. P. Dziobak, J. Kleissl, and D. L. Westphal, 2004 Regional and hemispheric impacts of anthropogenic and biomass burning emissions on summertime CO and O3 in the North Atlantic lower free troposphere, *J. Geophys. Res.*, **109**, D24310, doi:10.1029/2004JD005147.

[14] Hogan, T. F., and L. R. Brody, 1993 Sensitivity studies of the Navy's global forecast model parameterizations and evaluation of improvements to NOGAPS, *Mon. Wea. Rev.*, **121**(8), 2373-2395.

IOP Conf. Series: Earth and Environmental Science 7 (2009) 012008 doi:10.1088/1755-1307/7/1/012008

Aerosol modelling in MOCAGE and operational dust forecasting at Météo-France

M Martet[1] and V-H Peuch[2]

[1] Météo-France DP/Serv/Env. Toulouse, France
[2] CNRM Météo France. Toulouse, France

E-mail: maud.martet@meteo.fr

Abstract. MOCAGE is the multiscale 3D Chemistry-Transport Model of Météo-France. It is run operationally for Air Quality, UV and dust forecasting, daily up to 96h. Meteorological forcings are provided by our NWP suites, ARPEGE and ALADIN. Forecasts are uploaded to the French platform Prév'Air (http://www.prevair.org). MOCAGE is also a research tool, with over 25 publications, and it is used in several European projects, like GEMS or AMMA. Last, a specific version will become operational soon for emergency response in support of our responsibilities of RSMC and VAAC. In operations, three domains (two-ways nesting) are used : globe (2°), Europe (0.5°) and France (0.1°). On the vertical, MOCAGE extends from surface up to 5 hPa (L47) with hybrid coordinates (□-P). A semi-lagrangian scheme is used for advection, while turbulent diffusion and convection are parameterized, using the Louis and Bechtold schemes respectively. The representation of dusts is based upon a sectional approach with 5 bins. Dust emissions are computed using the scheme of Marticorena and Bergametti over Saharan and Chinese deserts. Wet deposition, sedimentation and dry deposition are also taken into account to compute concentrations. Various diagnostics are available daily: mass concentrations on different altitude levels, column, emissions, AOD and separate sink terms. We present an overview of the system and its validation studies.

1. Introduction

MOCAGE (Modèle de Chimie Atmosphérique à Grande Echelle) is the multiscale 3D Chemistry-Transport Model of Météo-France. This model is used in an operational way to forecast dust aerosol concentrations. In the following, a description of the model configuration used operationally will be developed, the forcings and the diagnostics of the operational configuration and then, some examples of the MOCAGE forecastings.

2. MOCAGE operational configuration

MOCAGE describes three imbricated domains in the operational configuration (figure 1):

- the Global domain, horizontal resolution 2°x2°,
- the European domain (12°W - 28°E ; 34°N - 62°N), horizontal resolution 0,5°x0,5°,
- the French domain (5°W – 10°E ; 41°N – 52°N), horizontal resolution 0,1°x0,1°.

© 2009 IOP Publishing Ltd

IOP Conf. Series: Earth and Environmental Science **7** (2009) 012008 doi:10.1088/1755-1307/7/1/012008

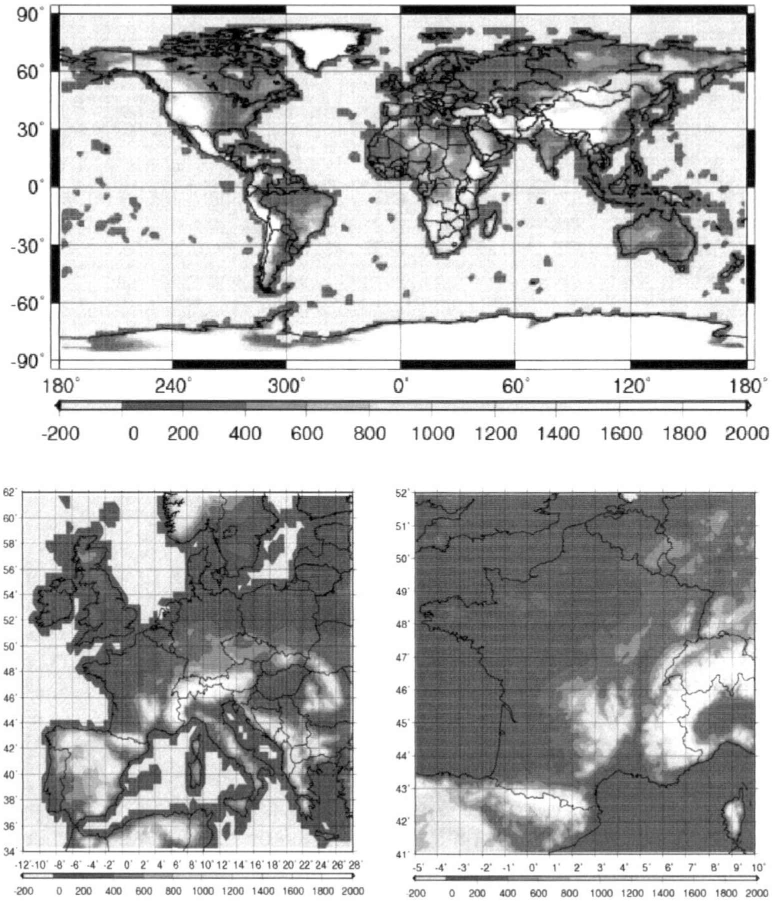

Figure 1: The three imbricated domains of the operational version of MOCAGE.

The communication between these domains is based on two-ways nesting. Vertically, 47 levels are considered in MOCAGE, from surface to 5hPa. The vertical coordinate is hybrid (σ-P), it follows the relief near the surface and the levels are at constant pressure at high altitudes.

The transport is ruled by a semi-lagrangian scheme for large-scale advection and parameterizations are included to subgrid-scale process: turbulent diffusion following [1] and convection is represented by a mass flux scheme described in [2].

To represent the evolution of aerosol concentration, some parameterizations specific to these particles have been introduced in MOCAGE:

- Wet deposition is composed by in-cloud and below-cloud scavenging. In cloud, the scavenging rate depends of the precipitation rate and the transfer efficiency of aerosol to droplets. Below cloud, the collision velocity and the collision efficiency are used to calculate the scavenging rate.
- The parameterization of dry deposition velocity [3] is composed of the aerodynamic resistance and the quasi-laminar resistance which contains Brownian diffusion, inertial impaction and inertial interception.

- The sedimentation velocity is calculated according to the Stokes law which takes into account non-continuum effects for small particles. A correction factor is introduced to consider the rarefaction of air at high altitudes.

In all these parameterizations, the particle size is important. So, the description of aerosol diameter has to be precise. We have chosen a sectional approach to represent aerosol distribution. For the operational configuration, the CPU time has to be reduced, that's why we have considered 5 bins. This scheme allows a good description of sinks following the aerosol size, especially sedimentation which is very important for desert dust.

To represent dust emissions, the module developed based on [4] and [5] has been introduced. Emissions are very precise and depend on meteorological conditions and ground characterizations. Emissions over Africa (12°N - 36°N; 17°W -78°E) and Asia (35.5°N - 47°N; 73°E - 125°E) are considered.

3. The operational forecastings

MOCAGE is a Chemistry-Transport model, so meteorological fields are needed to compute aerosol forecasts. Temperature, Humidity, Wind and Pressure fields come from Météo-France operational model ARPEGE and ALADIN. Everyday, a 96 hours forecast is performed and results are available on the internet via Prev'air site (http://www.prevair.org/fr/prevision_pous_desert.php). On this site, daily mean and maximum surface concentrations are shown for the next three days. But, more diagnostics are available on an intranet site and can be provided (figure 2):

- emitted quantities in (kg/m^2).
- concentrations at different levels (in kg/m^3) : surface, 500 m, 1000 m and 5000 m.
- total column of dusts concentrations (in kg/m^2).
- quantities eliminated by the sinks (in kg/m^2) : wet deposition, dry deposition, sedimentation.
- Aerosol optical thickness calculated using the Mie theory and considering that particles are spherical.

All these diagnostics are available on the three domains.

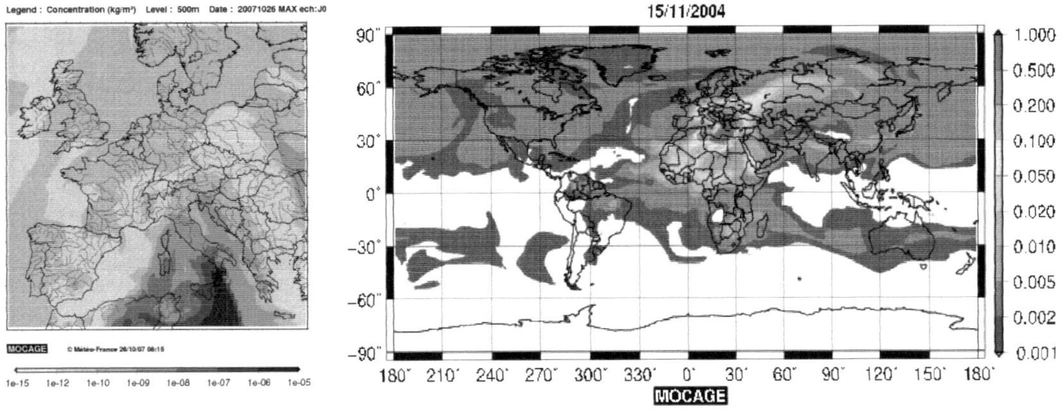

Figure 2. Two available diagnotics in MOCAGE forecasts : Dust concentration at 500 m on the European domain (left) and aerosol optical thickness on the Global domain (right).

4. Examples of MOCAGE results

In the following, two examples of MOCAGE forecasts are described. The first one happened in February 2004 and concerned South of France. The other one happened in May 2007, the dust cloud is lifted in West Africa, go through Atlantic ocean and has consequences on the visibility in Caribbean islands.

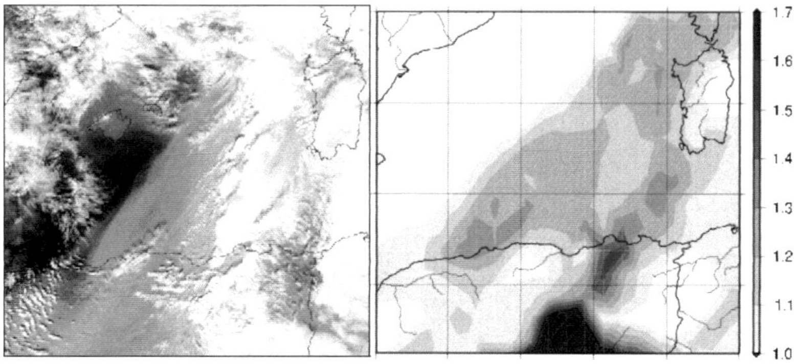

Figure 3. MODIS image (left) compared with MOCAGE dust total column in g/m^2 (right), February 21, 2004.

In February 2004, a dust cloud was lifted in North Africa and the satellite image shows it over the Mediterranean Sea. In the South of France, the PM10 observations recorded high values (more than 200 μg/m^3). This episode has been chosen to validate MOCAGE results. Qualitatively, the total column of dusts calculated by MOCAGE has been compared to the satellite image (figure 3). This comparison underlines that MOCAGE well represents the geographical position of the cloud over the Mediterranean sea. Quantitatively, PM10 measurements have been compared with dust surface concentration of MOCAGE (figure 4). The observations recorded two peaks with values higher than 200 μg/m^3. On MOCAGE results, the first peak is not represented. This problem is caused by meteorological forcings. This kind of dust transport is caused by a depression situated over the Mediterranean Sea. The difference in the position of the depression between the meteorological forecasts used in MOCAGE and the reality implies a non-realistic flux near the observation site. Considering the second peak, MOCAGE represents the end of the episode three hours later than the reality. Nevertheless, the results of MOCAGE are good: the timing and the intensity are well represented by our model.

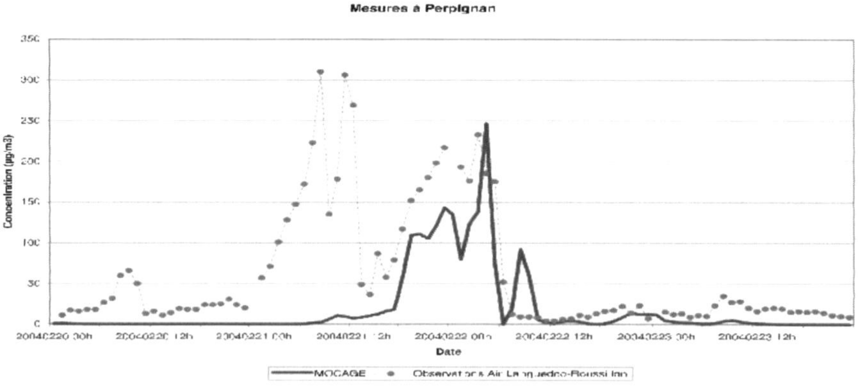

Figure 4. Episode of February 2004: comparison between PM10 measurements (red dots) and MOCAGE dust concentration (blue line) at Perpignan station.

In May 2007, a dust cloud was observed in West Africa. It is visible on MODIS image and a strong reduction of visibility has been observed in Martinique (French Caribbean island). Like in the case of February 2004, MOCAGE dust column has been compared to satellite image and shows a good representation of the dust cloud by MOCAGE during the travel of the cloud over the Atlantic ocean (figure 5). This comparison is qualitative and quantitative studies have to be performed. PM10 measurements in Martinique have recorded the highest value since 2000: 155 µg/m^3.

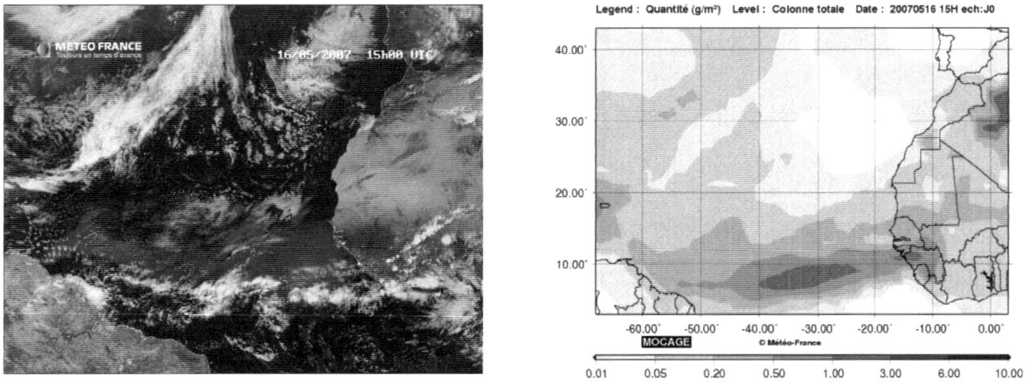

Figure 5. Multispectral colour-composite image (left) compared with MOCAGE dust total column in g/m^2 (right), May 16, 2007 1500 UTC.

Figure 6 shows the evolution of PM10 measured at Lamentin station and the desert dusts concentration calculated by MOCAGE.

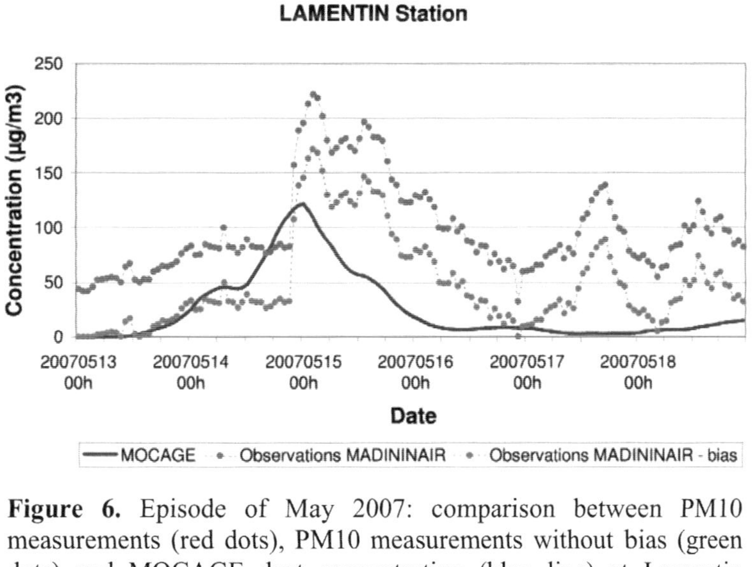

Figure 6. Episode of May 2007: comparison between PM10 measurements (red dots), PM10 measurements without bias (green dots) and MOCAGE dust concentration (blue line) at Lamentin station (Martinique).

The first day (13/05/2007), PM10 is about 50 µg/m^3 but the Saharan dust cloud has not reached Martinique yet. This concentration represents typical levels of PM10 at this monitoring site in the absence of any specific event. MOCAGE, in the configuration used here, does not account for local sources and these typical levels of pollutants cannot be represented for in the simulation. Thus, in order to compare MOCAGE simulation, we remove this background concentration of 50 μg/m^3. The dust concentration calculated by MOCAGE rises from 13/05/2007 12H00 UTC and reaches a maximum value of 121.8 μg/m^3 (15/05/2007, 00H00 UTC). During the two first days of simulations, observations and model data are in good agreement. MOCAGE represents the arrival of the dust cloud by a gradual increase of the concentration. The maximum value is modelled a few hours earlier than in the observations, but the maximum is in the right range. During the following days, MOCAGE evacuates the dust cloud more quickly than the reality. The concentrations are lower than 10 μg/m3 from 16/05/2007, 4H00 UTC in the model results and from 16/05/2007, 22H00 UTC in the observations. Moreover, two maxima are observed in the PM10 data (17/05/2007 17H00 UTC and 18/05/2007 13H00 UTC). These high concentrations are not represented in the model we think because of an underestimation of secondary Saharan dust during the days after the major emission.

5. Conclusion and perspectives

We have seen that our model is a performing tool to represent dust concentrations. The parameterizations introduced in MOCAGE allow us to use our model to operationally forecast dust concentrations. This model will also be used in AMMA simulations. Another capacity of MOCAGE is to represent back trajectories what could be useful to determine sources of dust from observations.

Other aerosol species will soon be implemented in MOCAGE : sea salts, volcanic ash and at medium range pollens. All these aerosols could have the same consequences on transport, health and air pollution.

References

[1] Louis J. F., 1979 A parametric model of vertical eddy fluxes in the atmosphere, *Bound.-Layer Meteor.*, **17**, 182-202.

[2] Bechtold P., E. Bazile, F. Guichard, P. Mascart and E. Richard, 2001 A mass flux convection scheme for regional and global models, *Q. J. R. Meteorol. Soc.*, **127**, 869-886.

[3] Nho-Kim, E.-Y., V.-H. Peuch, M.-J. Lee and S. N. Oh, 2004 Parameterization of size-dependent particle gravitational settling for global atmospheric transport modeling, *Atmospheric Environment*, **38**, 1933-1942.

[4] Marticorena B. and G. Bergametti, 1995 Modeling the atmospheric dust cycle : 1-Design of a soil derived dust production scheme, *J. Geophys. Res.*, **100**, 16415-16430.

[5] Marticorena B., G. Bergametti, B. Aumont, Y. Callot, C. N'Doumé and M. Legrand, 1997 Modeling the atmospheric dust cycle : 2-Simulations of Saharan dust sources, *J. Geophys. Res.*, **102**, 4387-4404.

IOP Conf. Series: Earth and Environmental Science **7** (2009) 012009 doi:10.1088/1755-1307/7/1/012009

Previsibility of Saharan dust events using the CHIMERE-DUST transport model

L Menut[1], I Chiapello[2] and C Moulin[3]

[1]Laboratoire de Météorologie Dynamique, Palaiseau, France
[2]Laboratoire d'Optique Atmospherique, CNRS/Universite des Sciences et Technologies de Lille, Villeneuve d'Ascq cedex, France
[3]Institut Pierre-Simon Laplace, Laboratoire des Sciences du Climat et de l'Environnement, CEA Orme des Merisiers, Gif sur Yvette, France

E-mail: menut@lmd.polytechnique.fr

Abstract. The previsibility of Northern Africa dust events is quantified using daily numerical forecast simulations for the next three days. The dust concentrations fields, modeled by CHIMERE-DUST, are first evaluate by comparisons to AERONET surface data, OMI and MSG Seviri satellite measurements. The accuracy and spread between measurements and model are discussed focussing on the first short observation period of the AMMA experiment in western Africa, with a simulation period ranging from January to March 2006. Second, the previsibility of dust is estimated by comparing model results for different leads in a forecats mode. The model skill is evaluate in term of capability to forecast (i) the surface wind speed (the key process for dust emissions), (ii) the dust emissions (depending on the wind speed as well as numerous others uncertain parameters, including threshold values on the friction velocity) and (iii) the transport of aerosols from source to remote areas (depending of horizontal transport, convection etc.). It is shown that emissions forecast can vary up to 80% (close to the sources) and that final forecasted dust concentrations and relative optical thickness do not exceed 40% and 20% in forecast variability.

1. The CHIMERE-DUST model

1.1. Scientific objectives

The main goals of the CHIMERE-DUST model are a better understanding and representation of emissions and long-range transport of thin dust layers. In this way, and due to the specific trends of mineral dust, it is important to validate the model with hourly simulations (to account for the small scales processes) but for long-term simulations (climatological studies). This is also a tool dedicated to quantify the amount of mineral dust transported from Africa to Western Europe for a better aerosol forecast at regional scale in the framework of European air quality analysis and forecast, such as with the PREVAIR french national forecast system [1]. In this way, the forecast is necessary and may also be considered as a good tool for research.

© 2009 IOP Publishing Ltd

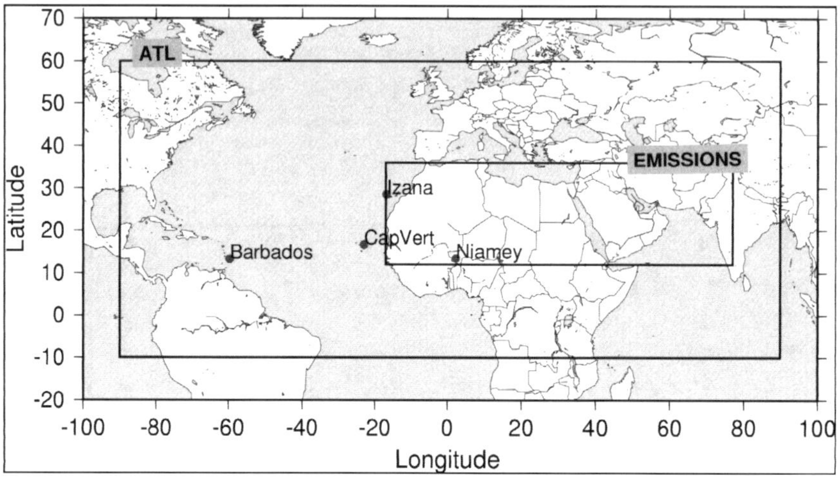

Figure 1. The model domain (ATL), with the emissions area (EMISSIONS). For the two, the horizontal resolution is 1°x1°.

1.2. Model characteristics

The model consists of three elements: (i) the meteorological platform with the MM5 model forced by the NCEP global meteorological fields, (ii) the dust emissions model and (iii) the CHIMERE-DUST transport model. These elements are used together and in the same manner both in analysis or forecast mode. CHIMERE-DUST is a specific version of CHIMERE and exists since 2005. The model was developed using modules from B.Marticorena, G.Bergametti and S.Alfaro (LISA), R.Vautard (LMD), among others.

The horizontal analysis data of NCEP [2] are provided on a regular 1.125°x 1.125°. These fields are used as boundary conditions and nudging for the MM5 model [3]. The dust emissions scheme used in the model is the [4] scheme. It computes horizontal fluxes from wind velocities and surface features for the emissions area (the 'EMISSIONS' area in the figure 1).

The dust vertical fluxes are computed by using the [5] parameterization, numerically optimized following [6]. The vertical fluxes are calculated corresponding to three dust size modes, then redistributed into the model size bins using a mass partition scheme. The wet deposition scheme is that described in [7]. The dry deposition velocity is parameterized following [8]. The transport model CHIMERE-DUST was developed on the basis of the chemistry-transport model CHIMERE [9,10]. The horizontal domain has an horizontal resolution of 1°x1° and is displayed as the "ATL" frame in figure 1. Vertically, 15 levels are defined from the surface to 200hPa. Turbulent parameters as u^*, the friction velocity and h the boundary layer depth are estimated from the mean meteorological parameters. The horizontal transport is performed using the Van Leer scheme [11] and the vertical mixing is estimated from the calculation of the bulk Richardson number as extensively described in [12]. The dust simulations are performed with a time-step of 7'30". A sectional aerosol size distributionis was chosen with 12 bins after [13] and [14].

2. Model "validation"

In order to validate the accuracy of the modelled dust concentrations, numerous periods were modelled and output concentrations were compared to available data such as AERONET network and SEVIRI/METEOSAT and MODIS satellite data. An example of comparisons between model and measurements is displayed in the figure 2 for the case of the 6 March 2004. The maps show comparisons of modelled dust load (g/m^2) for the dust plumes patterns validation. An example of

AERONET site comparison with the model for the validation of the hourly variability of surface dust concentrations is presented in terms of Aerosol Optical Thickness (AOT) for the Capo verde site.

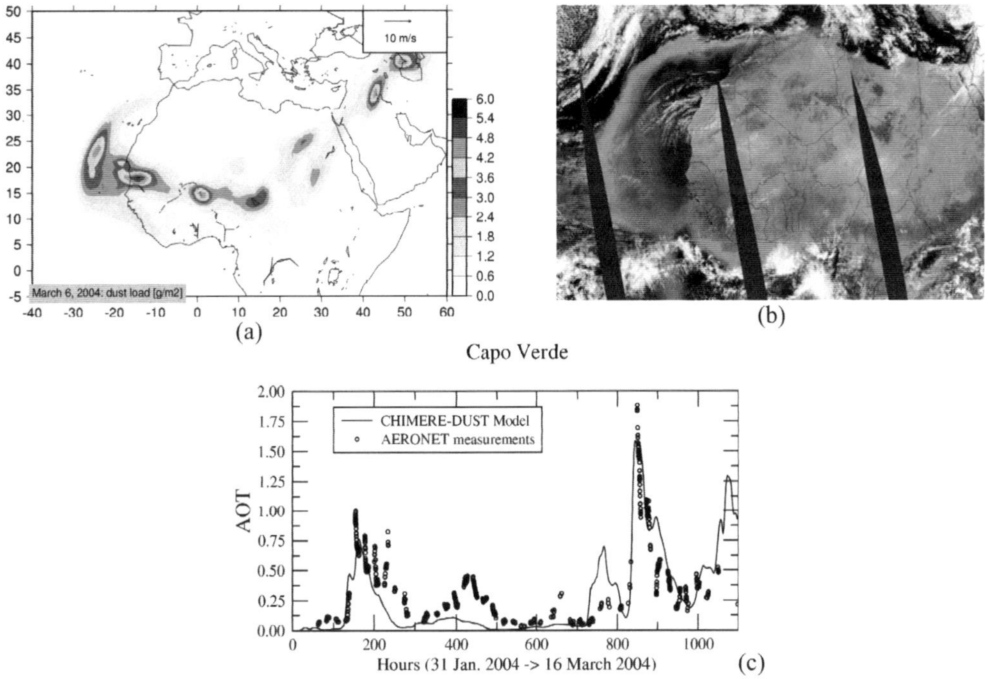

Figure 2. Dust load for the 6 March 2004 (a) CHIMERE-DUST model and (b) SEAWIFS; and (c) comparison of CHIMERE-DUST versus AERONET at Capo Verde site

3. The CHIMERE-DUST forecast

The forecast is done every day at the Laboratoire de Météorologie Dynamique in an experimental context since September 2005 and all output results are downloadable in free access via ftp. The model used for forecast is strictly the same than the version used for analysis or sensitivity studies [6,14]. The only diference is the meteorological forcing. In place of reanalysis fields, the NCEP forecast fields are downloaded every midnight and mesoscale MM5 runs are performed during the following night. After the meteorological runs, the CHIMERE-DUST used these fields to estimate dust emissions and transport. Forecasts fields are provided on a web site every morning for the next three days. The simulation is initialized the day before using the analysis meteorology and dust boundary conditions issued from the CHIMERE-DUST forecast done 24h before.

The home page of the experimental CHIMERE-DUST forecast is presented in the figure 3 (left). In parallel, output results are formatted to be directly compared to other available products on the web and for specific needs (such as during the AMMA experiment). Systematic comparisons are done with AERONET aerosol optical thicknesses (see for example figure 3 (right)) and with NAAPS model maps.

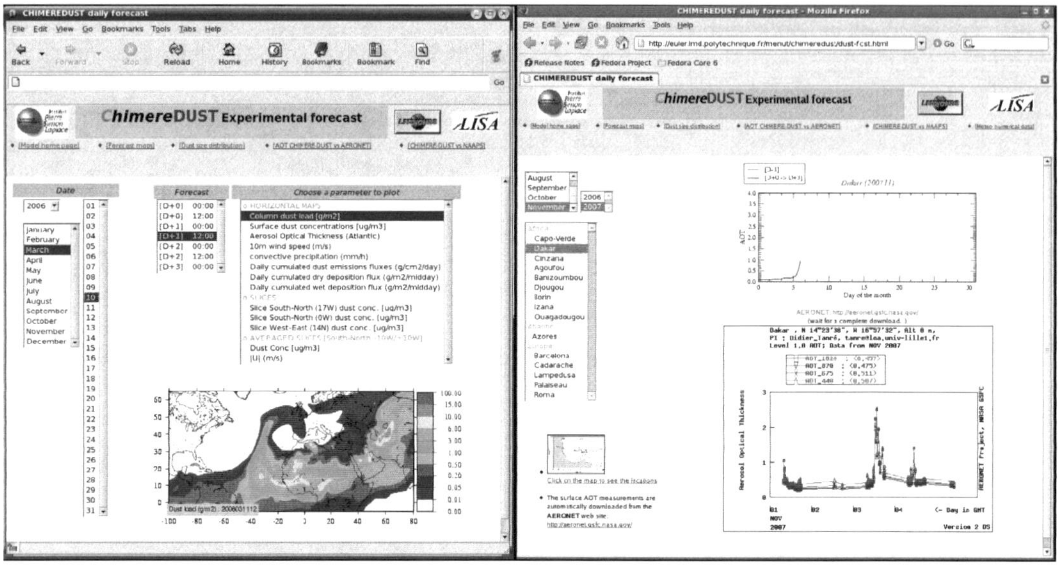

Figure 3. http://euler.lmd.polytechnique.fr/menut/chimeredust/dust-fcst.html. Daily comparison to AERONET level 1 AOT data (+ NAAPS model)

4. A specific test-case : The AMMA experiment

To illustrate the variability of the dust forecast, we present meteorological and dust concentrations fields estimated day after day and for different periods. The time period of interest correspond to the first three months of 2006 i.e the period of the First Short Observation Period of the African Monsoon Multidisplinary Analyses (AMMA) experiment (SOP0). The modelled results (MM5) are presented for the wind speed in the Figure 4 for two specific sites (Djougou and Dakar) and for the period from the 1st to the 10rd March 2006. Each hour, one can see one value for [D-1], [D+0], [D+1] and [D+2], where [D-1] is the "analysis" and the others days are forecast.

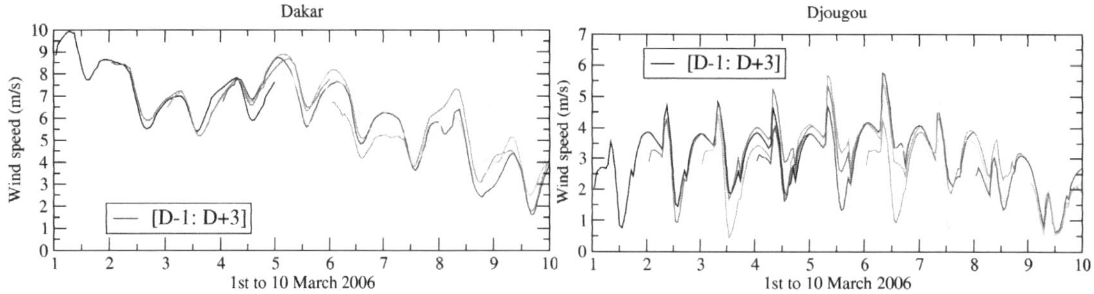

Figure 4. Time series of forecasted wind speed (m/s) for Djougou and Dakar, from the 1st to the 10rd March 2006. For each hour, the several leads are superimposed.

As presented in the figure 4, the day to day wind speed variability may be important as ±2m/s. For low values, typically less than 7 m/s when dust are never emitted, this variability is not an important factor of uncertainty for dust forecast: in this case, the error is only linked to a change in the plume transportation and potentially its direction. Around the saltation friction velocity threshold (the fraction velocity used to estimate dust emission or not), this variability may be very important: as a function of the forecast wind speed value, the dust emission model may diagnose dust fluxes or not at all. Thus, a real uncertainty exists around this threshold value and the next point to study is to evaluate if this has an impact on dust emission over the whole domain.

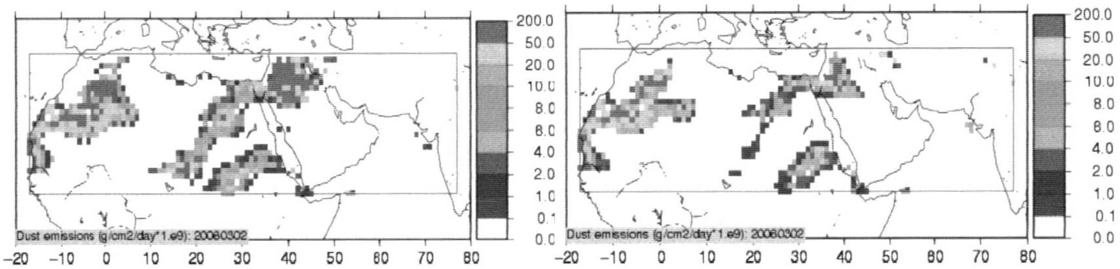

Figure 5. Surface dust emissions fluxes maps. The red frame represents the whole emissions domain. Dust fluxes are summed up over the day and presented in g/cm²/day.

Figure 5 presents dust emissions fluxes calculated for the same time (1st march 2006, 12:00 UTC) but with two different meteorological forcings: the analysis (i.e the more accurate fields) and the forecast for two days later (i.e computed on 3 March 2006). The impact of a small error on wind speed clearly may induce a large error on emitted dust: at the center of the domain, important fluxes are diagnosed two days in advance. Using the analyszed wind field, these fluxes are removed. This is a case where non existing dust plume may be forecast.

Using these dust emissions fluxes, the CHIMERE-DUST model runs in forecast mode and estimates the transport and deposition of dust concentrations. Vertically integrating the whole atmospheric column of dust concentrations, the dust load may be estimated as well as the aerosol optical thickness. These two quantities are directly compared to the OMI aerosol index and the SEVIRI AOT, as presented in the figure 6. With OMI, there is no direct estimation of AOT but the signal may be use over ocean and over land. Comparing the plumes horizontal structures between OMI and CHIMERE-DUST, we can see a very good representation of the dust plume by the model. On the other hand, SEVIRI is able to estimate an accurate AOT but only over the ocean. The comparison with the model also shows that the magnitude of the modelled AOT is of the same order of magnitude than the satellite retrieval. With these two types of information, we can conclude than the model is able to reproduce both intensities and patterns of dust plumes.

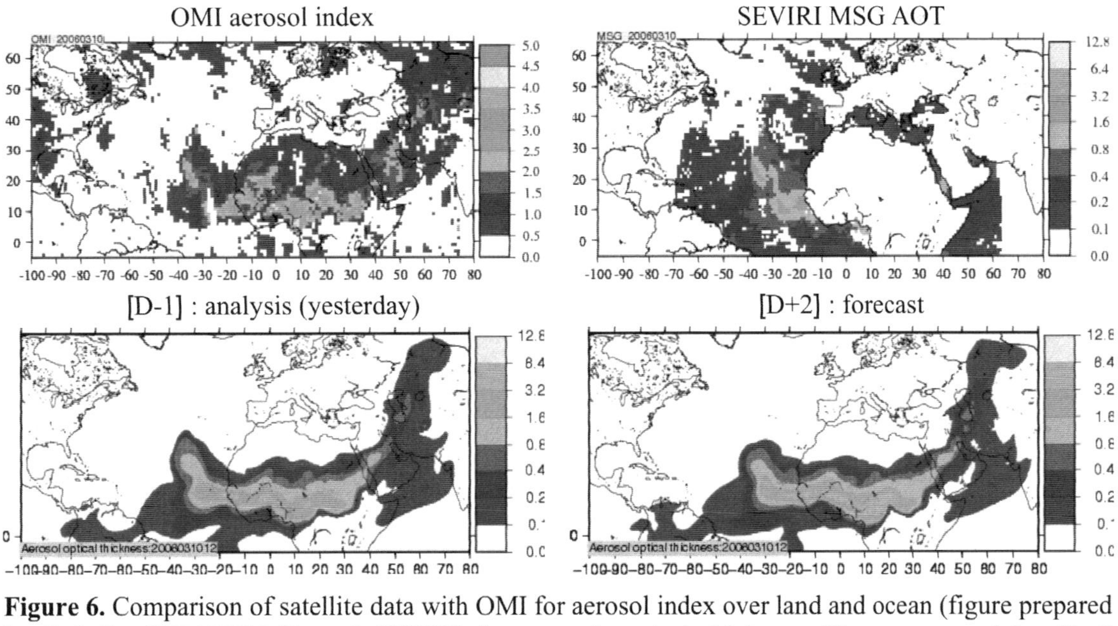

Figure 6. Comparison of satellite data with OMI for aerosol index over land and ocean (figure prepared by Isabelle CHIAPELLO) and SEVIRI for aerosol optical thickness (figure prepared by Cyril MOULIN) and modelled aerosols optical thickness for the 10 March 2006, 12:00 UTC.

To go further in the comparison, we can directly analyze dust surface concentrations. This quantity is of great interest in the framework of air quality studies and an accurate forecast would be of great importance for the global particulate matter forecast over urbanized areas for example. In the figure 7, we compare the modelled surface concentrations for several sites and several leads. The comparison to data is not presented here because data are a global budget of many particulate matter origins (dust, biomass burning, resuspension of locally produced matter etc.) and thus the comparison modelled dust to observed particulate matter has no sense.

The results are presented in the figure 7 for the 10 March 2006 and for the sites of Banizoumbou and Djougou in Africa (near the sources), Roma and Barcelona (in Italy and Spain, respectively and acting as remote sites). The [D-1] time series may be considered as the 'reference' case since this is the simulation with the 'best' (i.e analyzed) meteorology. Near the sources, in Banizoumbou and Djougou, the modelled concentrations are huge (several hundred μg/m^3). The temporal variability is the same between every leads but the magnitude of the concentrations varies a lot. It is not possible to estimate a systematic bias or a systematic tendency correlated to the lead: depending on site and period, the modelled concentrations may be under- or over-estimated compared to the reference. But a more general result may be expressed: a variability up to 50% is observed between the [D-1] and the [D+1] simulations. Over Europe, the concentrations are really lower and in accordance with usually recorded values in these regions. The variability is in general of the same order of magnitude than over the near sources sites and may reach ±50%. This variability is important to quantify for the uncertainty estimation of air quality forecast in Europe.

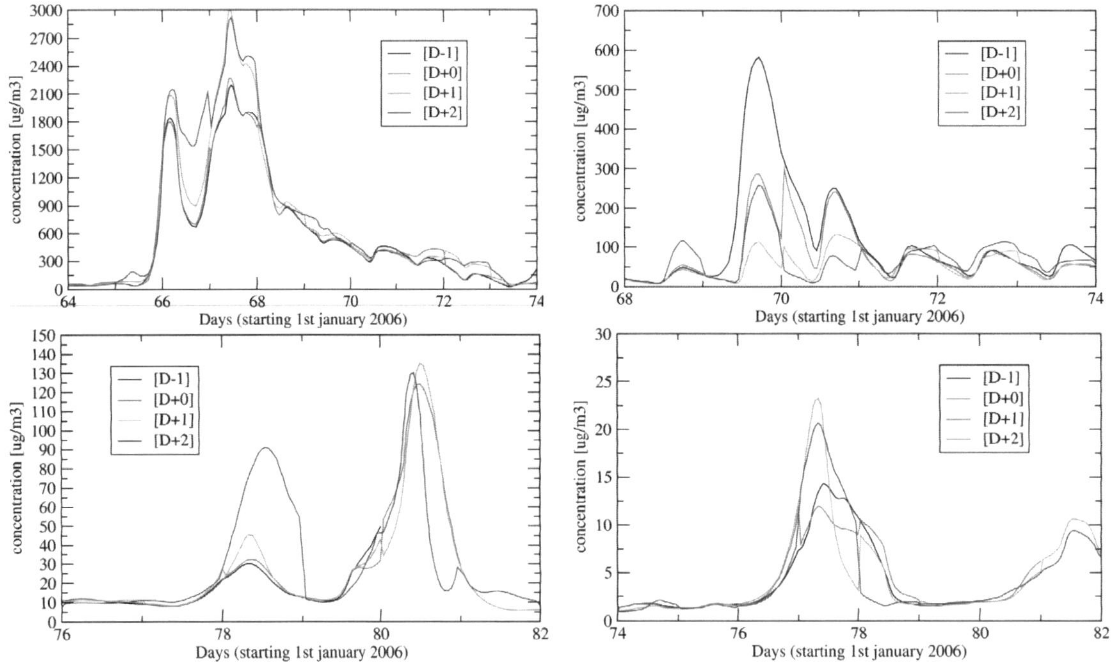

Figure 7. Dust concentrations time series for several sites in Africa (Banizoumbou and Djougou) and western Europe (Roma and Barcelona). For each time, all forecast leads are reported.

5. Conclusion

Every day, the CHIMERE-DUST model forecasts four days of mineral dust emissions and transport since 2005. The results are displayed every day on a dedicated web plat-form, including maps, time series, comparisons to observations and to other models. The output data are also provided on the web site for researchers interested to work with.

In this paper, the variability of the dust forecast was extensively studied by comparing model and measurements during the first three months of 2006 and not only for the most important dust events: they correspond to the most easy periods to model with a dust model since the potential error on wind speed becomes negligible compared to the threshold friction velocity. The model was used for all days of the periods including non dusty days: with winds close to the estimated threshold, these days are the more difficult to model and the model has to be enough accurate to avoid false dust emissions and thus false forecasts of dust plume transport. The model was systematically compared to available observations (AERONET stations, satellites) and some comparisons were presented. For each studied period, the variability of the forecast was shown and the comparisons shows good results for the large scale patterns. The AOT are close and the dust plumes leaving the western Africa and transport towards Europe are well forecast in size and intensity. Thus we can conclude that the forecast is relatively stable at the continental scale. Far from the sources and in Europe, the forecast shows its sensitivity to all model errors made from the emissions to the transport. Between the analyzed day and the forecast for two days later, the dust concentrations differences ranged from 0 to 50%. This shows that, for an operational use for air quality forecast over Europe, a non negligible uncertainty remain in the modelling of Saharan dust emission and transport.

References

[1] Honoré, C. et al., 2008 Predictability of european air quality: The assessment of three years of operational forecasts and analyses by the prev'air system, J. Geophys, Atmos, in press.

[2] Kalnay, E., et al. 1996 The ncep/ncar 40-year reanalysis project, Bull. Amer. Meteorol. Soc., pp. 437-471.

[3] Dudhia, J. 1993 A nonhydrostatic version of the penn state/ncar mesoscale model : validation tests and simulation of an atlantic cyclone and cold front., Mon. Wea. Rev., 121, 1493-1513.

[4] Marticorena, B., and G. Bergametti 1995 Modeling the atmospheric dust cycle: 1 Design of a soil derived dust production scheme, J. Geophys. Res., 100, 16,415-16,430.

[5] Alfaro, S. C., and L. Gomes 2001 Modeling mineral aerosol production by wind erosion: Emission intensities and aerosol size distribution in source areas, J. Geophys. Res., 106, 18,075-18,084.

[6] Menut, L., C. Schmechtig, and B. Marticorena 2005 Sensitivity of the sandblasting fluxes calculations to the soil size distribution accuracy, J. Atmos. Ocean. Tech, 22, No. 12, 1875-1884.

[7] Loosmore, G., and R. Cederwall 2004 Precipitation scavenging of atmospheric aerosols for emergency response applications: testing an updated model with new real-time data, Atmos. Env., 38, 993-1003.

[8] Venkatram, A., and J. Pleim 1999 The electrical analogy does not apply to modeling dry deposition of particles, Atmos. Env., 33, 3075-3076.

[9] Vautard, R., M. Beekmann, J. Roux, and D. Gombert 2001 Validation of a hybrid forecasting system for the ozone concentrations over the Paris area, Atm. Env., 35, 2449-2461.

[10] Bessagnet, B., A. Hodzic, R. Vautard, M. Beekmann, S. Cheinet, C. Honoré, C. Liousse, and L. Rouil 2004 Aerosol modeling with chimere : preliminary evaluation at the continental scale, Atm. Env., 38, 2803-2817.

[11] Van Leer, B. 1979 Towards the ultimate conservative difference scheme. V A second order sequel to Godunov's method, J. Computational Phys., 32,101-136.

[12] Menut, L. 2003 Adjoint modelling for atmospheric pollution processes sensitivity at regional scale during the ESQUIF IOP2, J. Geophys. Res., 108 (D17), 8562, doi :10.1029/2002JD002549.

[13] Forêt, G., G. Bergametti, F. Dulac, and L. Menut 2006 An optimized particle size bin scheme for modeling mineral dust aerosols, J. Geophys. Res., 111, D17310, doi:10.1029/2005JD006797.

[14] Menut, L., G. Forêt, and G. Bergametti 2007 Sensitivity of mineral dust concentrations to the model size distribution accuracy, J. Geophys. Res., 112, D10210, doi:10.1029/2006JD007766.

Dust forecasting system in JMA

M Mikami[1], T Maki[2] and T Y Tanaka[1]

[1] Meteorological Research Institute, Japan Meteorological Agency. Tsukuba, Japan.
[2] Numerical Prediction Division, Japan Meteorological Agency. Tokyo, Japan.

Email: mmikami@mri-jma.go.jp

Abstract. JMAs dust forecasting information, which is based on a GCM dust model, is presented through the JMA website coupled with nowcast information. The website was updated recently and JMA and MOE joint "KOSA" website was open from April 2008. Data assimilation technique will be introduced for improvement of the "KOSA" information.

1. JMA dust forecast model and system

JMA started to operate an aeolian dust information service through its website (http://www.jma.go.jp/en/kosafcst/index.html) from January 2004. Both prediction and observation information, which are shown with a map of East Asian region, are presented there. The observation information is based on the present weather code and visibility obtained from SYNOP reports in the region from 20°N to 50°N and from 110°E to 150°E.

For the prediction of Aeolian dust, we developed a GCM dust model, MASINGAR (Model of Aerosol Species IN the Global Atmosphere; [1]). This model was developed based on the chemical transport model in MRI/JMA98 and is composed of four different physical processes of Aeolian dust, (1) wind erosion scheme, (2) boundary layer module, (3) long-range transport model, and (4) wet and dry deposition scheme (figure 1).

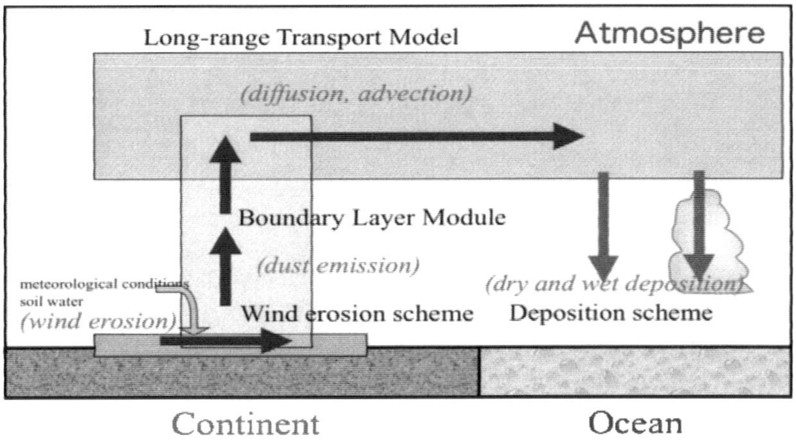

Figure 1. Schematic diagram of MASINGAR

© 2009 IOP Publishing Ltd

Figure 2 shows the model structure of MASINGAR. The dust module is directly coupled with the general circulation model (GCM). For the better representation of size distribution, we have divided the dust particle size into 10 bins from 0.2 to 20 μm diameter (Table 1).

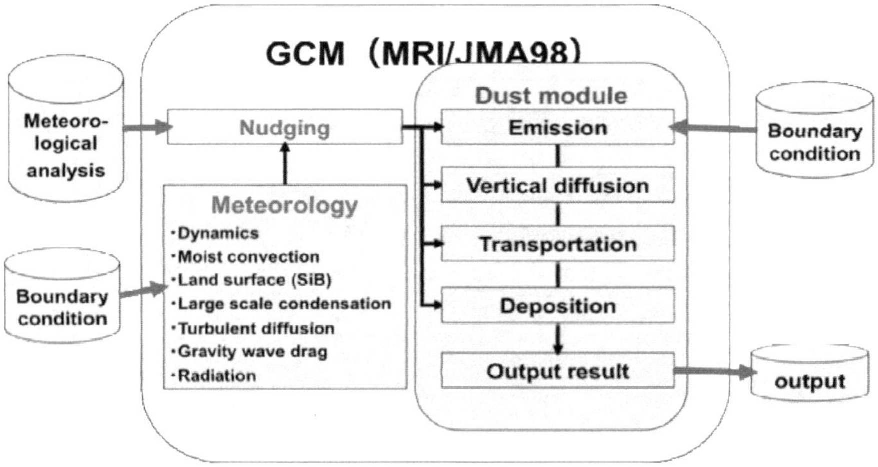

Figure 2. The model structure and flow of MASINGAR.

Table 1. Specifications of GCM aerosol model (MRI/JMA98AGCM) and GCM dust model (MASINGAR).

Atmospheric Model	**MRI/JMA98AGCM [2]**
Dynamics	General Circulation Model (Spectral Model)
Resolution	T106 (1.125°), L30 (~0.4hPa)
Time Integration	Semi-implicit Scheme
Cumulus Convection	Arakawa-Schbert Scheme
Radiation	Shotwave: delta-two-stream approximation Longwave: multi-parameter random model
Surface Processes	SiB
Turbulent Diffusion	Mellor-Yamada (Level 2 Closure)
Nudging	Meteorological analysis, forecast, and snow depth analysis
Dust module	**MASINGAR [1,3]**
Dust bin size	10 bins (0.2~20μm)
Emission	Gillette Scheme
Transport	3D-Semi Lagrangian Scheme
Diffusion	Turbulent and cumulus convection
Deposition	Dry and Wet Deposition

As well known, dust emission flux depends on the friction velocity, which expresses the strength of shear stress, and is formulated as a function of friction velocity. JMA dust model, however, uses the simpler formula applying wind speed near the ground surface (10m) for the simplicity of the routine calculation. Then the dust emission flux is expressed as (equation 1):

$$F=C\cdot M\cdot A\frac{W_{gt}-W_{g}}{W_{gt}}\left(U_{10}-U_{t}\right)U_{10}^{2}\quad \text{for}\quad U_{10}>U_{t} \tag{1}$$

where F is dust emission flux, C is dimensional factor, M is mass distribution, A is erodible fraction, W_g is soil moisture, W_{gt} is threshold value of W_g (=0.3 kg/m^2), U_{10} is wind velocity at 10m, and U_t is threshold wind speed (=6.5 m/s at 10m).

2. "KOSA" information service by JMA (15 Jan. 2004 to 26 Feb. 2008)

The word "KOSA" is named after an Aeolian dust event in East Asia, being a Japanese word originally well known in Japanese. For this, the Japanese version of JMA's website is called as "KOSA" information service. The contents of "KOSA" information service by JMA are provision of past to present surface observation (Japanese/English), "KOSA" warning information (Japanese only), and "KOSA" prediction (Japanese/English) (figure 3). The English version of the prediction information is named as Aeolian dust prediction instead of "KOSA".

- Past to present surface observation: Figure 3 shows an example of the information system. Marked in the figure are the weather stations in East Asia (20º-50º N, 110º-150º E), which observed Aeolian dust or local sand/dust haze during that day. The figure is updated whenever new data become available.
- "KOSA" information: When a dust event is observed or predicted by the model, JMA provides the information via the website. On average, there are 1,000,000 hits/event on the website.
- "KOSA" prediction: When a dust event (dust storm, suspended dust) is reported in East Asia, the predicted KOSA region (yellow color in the figure that indicates visibility less than 10 km) is shown on the website. This information is updated at around 5 am (JST) every day.

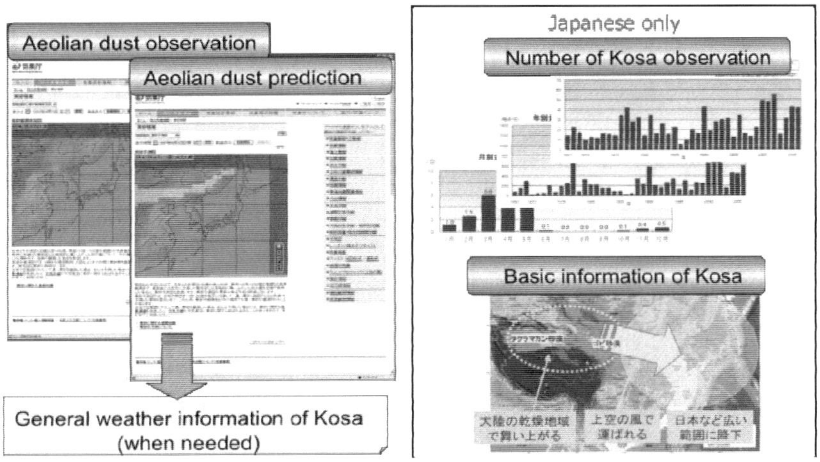

Figure 3. Images of "KOSA" information service by JMA (until 26 Feb 2008)

3. Model improvements

We improved the model concerning the following points:

- We updated the vegetation map of the Gobi Desert, northeastern China and Nei-Mongol based on MODIS 2000-2001 data and JMA's analysis of Aeolian dust observations in 2006 (figure 4).
- We improved the nudging procedure with JMA operational analysis and forecast (T213L40) data.

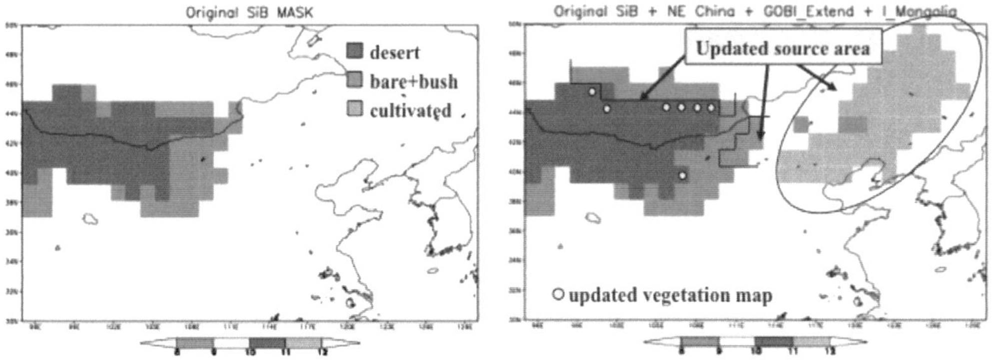

Figure 4. Original source area (left) and updated source area (right) of MASINGAR

- We validated the new boundary conditions in the improved model version against the surface weather observations in Japan using Threat Score (TS, Table 2), which is defined as (equation 2):

$$TS = \frac{FO}{FO + FX + XO} \tag{2}$$

where FO is the number of events which were predicted by "KOSA" and verified by observations according to table 2 (the same scheme applies for FX and XO)

Table 2. Threat Score categories and criterion.

KOSA event Forecast	
Category	Surface* dust concentration
KOSA Predicted (F)	More than 90 $\mu g/m^3$
Not Predicted (X)	Less than 90 $\mu g/m^3$

* Surface: 0-1 km

KOSA event Observation	
Category	Present Weather (ww)
Event Observed	06-09, 30-35,98
No Event Observed	00-05
Unknown	The others

The results indicated that the Threat Scores are improved by updated boundary conditions (table 3). Figure 5 shows an example of the 48 hours and 120 hours forecast results. Even 120 hour forecast shows reasonable result.

Table 3. Threat score of MASINGAR against surface observation in Japan.

Forecast Period	**00-24h**	**24-48h**	**48-72h**	**72-96h**
2006 (Old model)	0.171	0.163	0.172	0.171
2007 (New model)	0.524	0.524	0.491	0.429

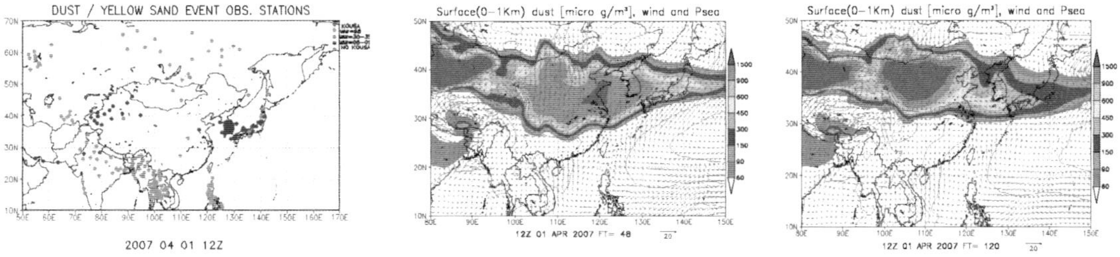

Surface Observation 48 hour forecast 120 hour forecast

Figure 5. An example of 48 and 120 hours KOSA forecast by MASINGAR (1 April 2007).

4. Updated "KOSA" information service

From 27 Feb. 2008, JMA updated the "KOSA" information service. The key improvements are the following:

- We extended the forecast period from 48 hours to 96 hours (figure 6 (1)).
- Predicted dust concentration at surface and column dust amount are colored in terms of its concentration (figure 6 (2)).
- Nowcasting information (past and present surface weather map) is also colored including the visibility (figure 6 (3)).

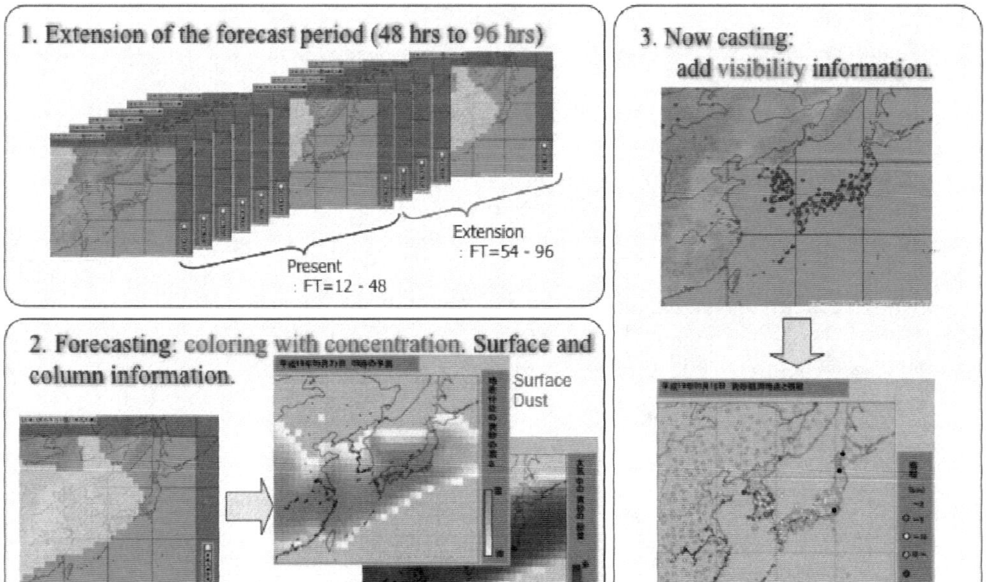

Figure 6. Updated information of KOSA web site by JMA.

5. Improved presentation and future work

From April 17 2008, the Ministry of Environment (MOE) and JMA started to operate a joint "KOSA" website (http://www.data.kishou.go.jp/obs-env/kosateikyou/kosa.html). From this homepage, we can move to the following linked pages: (1) Nowcast information of meteorological stations (JMA), (2) nowcast information of remote sensing site (MOE), (3) forecast information (JMA), (4) SPM (Suspended Particulate Matter) information (MOE), (5) PM10 (Particulate Matter less than 10 micron diameter) information (link to KMA, Korea Meteorological Administration), and (6) statistical information of KOSA (JMA).

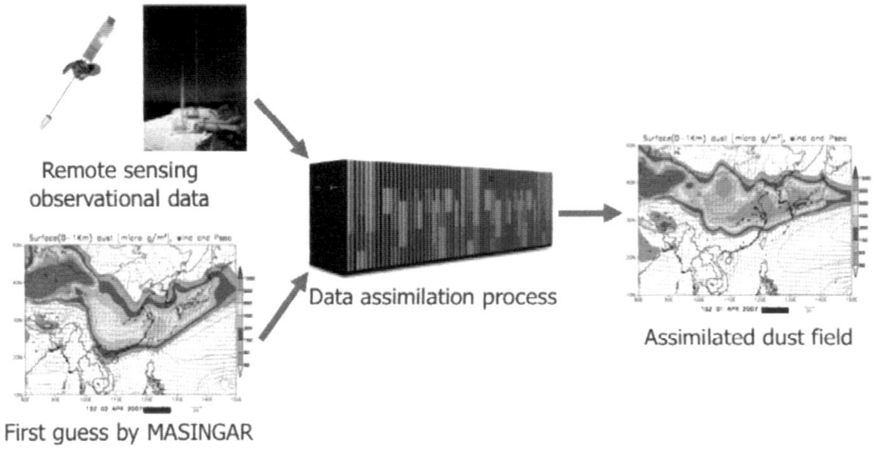

Figure 7. Schematics of the data assimilation of KOSA forecast by MASINGAR

We are also planning to introduce a data assimilation technique in order to improve the model performance. For this, we will apply an inverse method for the better representation of dust flux from source regions using remote sensing data (satellite, lidar etc.). Figure 7 shows the scheme for data assimilation.

We see two different main tasks for the future development of dust models; one is focused on a better dust forecast and the other one is related to dust research. For the model improvements, we need to develop data assimilation techniques in order to get a better adjustment with real time data. For this, collaboration with MOE and Asian countries are required. A better horizontal resolution of the model (110 km at present) and up-to-date GIS data set are also required.

On the other hand, for dust research, more process studies on dust emission, deposition (wet and dry), and dust particle characterization during long-range transport are desirable. For the better understanding of dust-climate interaction, researches on direct/indirect aerosol radiative perturbations related to dust including dust-cirrus formation and dust albedo effects on snow surface are also required. Intensive observations and experiments coupled with model work are required for this.

References

[1] Tanaka, T. Y., Orito, K., Sekiyama, T. T., Shibata, K., Chiba, M., and Tanaka, H., 2003 MASINGAR, a global tropospheric aerosol chemical transport model coupled with MRI/JMA98 GCM: Model description. *Pap. Meteor. Geophys.* **53**, 119–138.

[2] Shibata, K., H. Yoshimura, M. Ohizumi, M. Hosaka, and M. Sugi, 1999 A simulation of troposphere, stratosphere and mesosphere with an MRI/JMA98 GCM. *Pap. Meteor. Geophys.* **50**, 15–53.

[3] Tanaka, T. Y., and Chiba, M., 2005 Global simulation of dust aerosol with a chemical transport model, MASINGAR. *J. Met. Soc. Japan.* **83A**, 255-278.

A Unified Atmospheric Model Suitable for Studying Transport of Mineral Aerosols From Meso to Global Scales

Z Janjic, H Huang and S Lu

National Centers for Environmental Prediction. Camp Springs, Maryland. USA

E-mail: Zavisa.Janjic@noaa.gov

Abstract. Highlights of current dust and mineral aerosol modeling activities at the National Centers for Environmental Prediction (NCEP), and projections for the future are presented. On the atmospheric modeling side, current activities are reviewed on the development of a unified nonhydrostatic atmospheric model that can be used for driving dust and aerosol models on a wide range of scales extending from meso to global. Finally, some of the outstanding problems associated with modeling of the transport of atmospheric constituents are discussed.

1. Highlights of NCEP dust and aerosol modeling progress and plans

The system that is currently being developed relies on the NCEP spectral Global Forecasting System (GFS) (http://www.emc.ncep.noaa.gov/gmb/moorthi/gam.html) for generating the atmospheric fields to drive the NASA Goddard GOCART model [1] handling the dust, mineral aerosols and chemical constituents. In this basic system the GOCART model is only passively driven by the GFS, without interacting with the driving model. Dust modeling only is included in the first step, and later on more aerosols will be added.

The main goals of the described base system are:

- To serve as an initial benchmark for future more advanced, fully integrated systems with two way interaction between the atmospheric model and the aerosol/chemistry,
- To provide aerosol fields for lateral boundary conditions for the regional air quality forecasting over North America; for this purpose the NCEP regional North American Mesoscale model (NAM) driving the Environmental Protection Agency (EPA) CMAQ air quality model (http://www.epa.gov/AMD/CMAQ/index.html) is used,
- To provide initial first guess aerosol fields for the data assimilation development, and,
- To meet the global dust forecasting goals of the National Weather Service (NWS) in support of the dust forecasting program of the World Meteorological Organization (WMO).

The verification of the dust/aerosol forecasts will be made using atmospheric optical depth (AOD) data and vertical profile measurements where available.

Some results from the GFS-GOCART simulation for the period of 12/27/07 to 12/30/07 (Asian dust storm) with the GFS using T126 resolution can be seen in figure 1. The values shown are the daily mean surface-layer concentrations for the bin size of 1.0 to 1.8 μm in $\mu g/m^3$. More results obtained

© 2009 IOP Publishing Ltd

using the current system can be found at http://www.emc.ncep.noaa.gov/mmb/hchuang/web/html/dust_day.html.

The near future plans include completion of the base system, its evaluation for the summer of 2008, and validation of the lateral boundary conditions provided by the system for use in regional air quality forecasting. The US Navy Dust/Smoke emission source estimates will be evaluated and possibly improved. As the beginning of the transition to the next phase, the Earth System Modeling Framework (ESMF) I/O couplers (http://www.esmf.ucar.edu/) will be added to the GFS and a new Nonhydrostatic *Multiscale* Model on the B grid (NMM-B, [2], [3]) that is being developed at NCEP and that will be reviewed in the next section. In addition, the assimilation of satellite AOD and radiance data will commence.

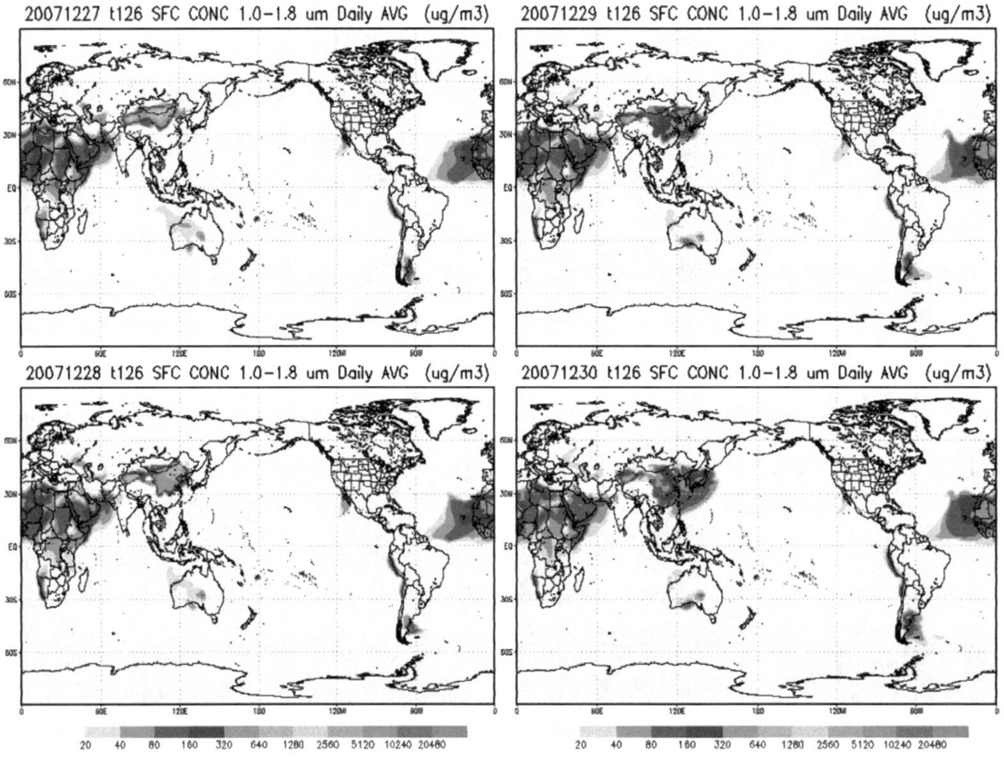

Figure 1. The GFS-GOCART simulation for the period of 12/27/07 to 12/30/07 (Asian dust storm) with GFS T126 (~1°x1°) resolution. The values shown are the daily mean surface-layer concentrations for the bin size of 1.0 to 1.8 μm in $\mu g/m^3$.

In the next, ESMF phase, transition to, and testing of fully integrated, two-way interactive systems is envisaged, consisting of both the GFS and NMM-B combined with GOCART. Extension of the systems to accommodate wider variety of aerosols is also expected. An example of the benefits of using the simulated aerosol fields in the GFS instead of climatological aerosol distribution is presented in figure 2. The bias and RMS errors of 24 hour (black lines) and 48 hour (red lines) forecasts of temperature profiles verified against RAOBs during a 2-month experiment with T126L64 resolution are shown for the runs with climatological aerosols (solid lines) and simulated aerosols (dashed lines). As can be seen from the figure, a more realistic simulated distribution of aerosols reduces temperature errors of both 24 hour and 48 hour forecasts.

In addition to the work at NCEP, a regional dust forecasting system devoted primarily to health issues is being developed and tested at University of Arizona, Tucson. The University of Arizona is acting as an associated research center. This system is based on the DREAM dust model [4] that is driven by the old NCEP Eta model [5,6,7,8,9,10]. Currently, transition is under way from the Eta to the new NCEP Nonhydrostatic *Mesoscale* Model (WRF NMM, [11], [12]). Possible future operational application of this system will be through a Regional Specialized Meteorological Center (RSMC).

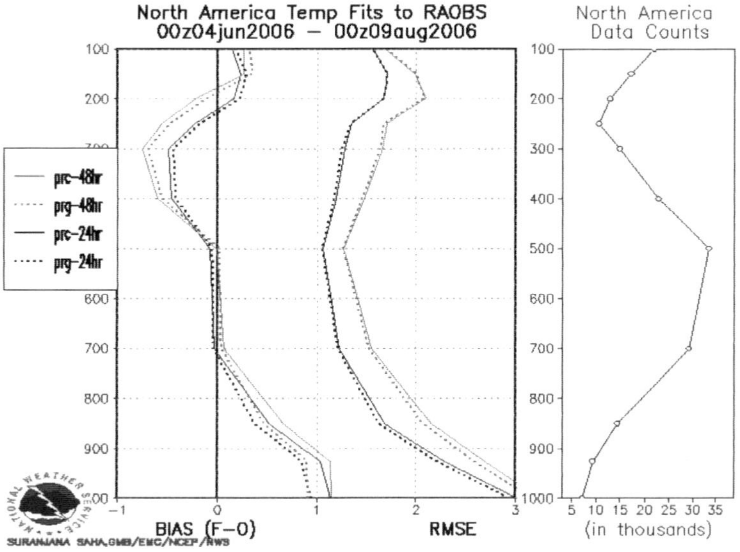

Figure 2. The bias and RMS errors of 24 hour (black lines) and 48 hour (red lines) forecasts of temperature profiles (left panel) verified against RAOBs during a 2-month T126L64 global data assimilation experiment using the GFS for the runs with climatological aerosols (solid lines) and simulated aerosols (dashed lines).

2. NMM-B, a new NCEP multiscale model

The NMM-B [2,3] is a new unified atmospheric model for a broad range of spatial and temporal scales. The NMM-B is being developed within the Earth System Modeling Framework (ESMF) at the National Centers for Environmental Prediction (NCEP) following the general modeling philosophy of the NCEP's WRF NMM [5,6] which is operationally used at NCEP as the regional NAM model. However, in contrast to the NMM, which is defined on the Arakawa E grid, the dynamics of the NMM-B has been reformulated for the Arakawa B grid in the (rotated) longitude-latitude coordinate system. The nonhydrostatic component of the model dynamics is introduced through an add–on module that can be turned on or off. The extra computational cost of the nonhydrostatic dynamics is low, or nonexistent if the nonhydrostatic extension is switched off at coarser resolutions.

The "isotropic" horizontal differencing employed in the model conserves a variety of basic and derived dynamical and quadratic quantities and preserves some important properties of differential operators. Among these, the conservation of energy and enstrophy improves the accuracy of the nonlinear dynamics of the model on all scales.

"Across the pole" polar boundary conditions are specified in the global limit. The polar filter selectively reduces tendencies, and thus decelerates, the wave components of the basic dynamical variables that would otherwise propagate faster in the zonal direction than the waves of the same wave length propagating in the meridional direction. In regional applications the rotated longitude-latitude system is used. With the Equator of the rotated system running through the middle of the integration domain, more uniform grid distances are obtained. In the vertical, the hybrid pressure-sigma coordinate is used. The forward-backward scheme is used for horizontally propagating fast waves, and an implicit scheme is applied for vertically propagating sound waves. The Adams-Bashforth scheme is employed for horizontal advection of the basic dynamical variables and for the Coriolis force. In order to eliminate stability problems due to thin vertical layers, the Crank-Nicholson scheme is used to compute the contributions of vertical advection.

A variety of WRF physical parameterizations have been coupled to the model. This variety is expected to be further extended in the future.

The guiding coding principles have been simplicity, transparency and the ESMF framework. The ESMF is expected to facilitate collaboration with community primarily through I/O compatibility. Even though the ESMF adds some extra complexity, the overhead remains reasonable compared to other modeling infrastructures.

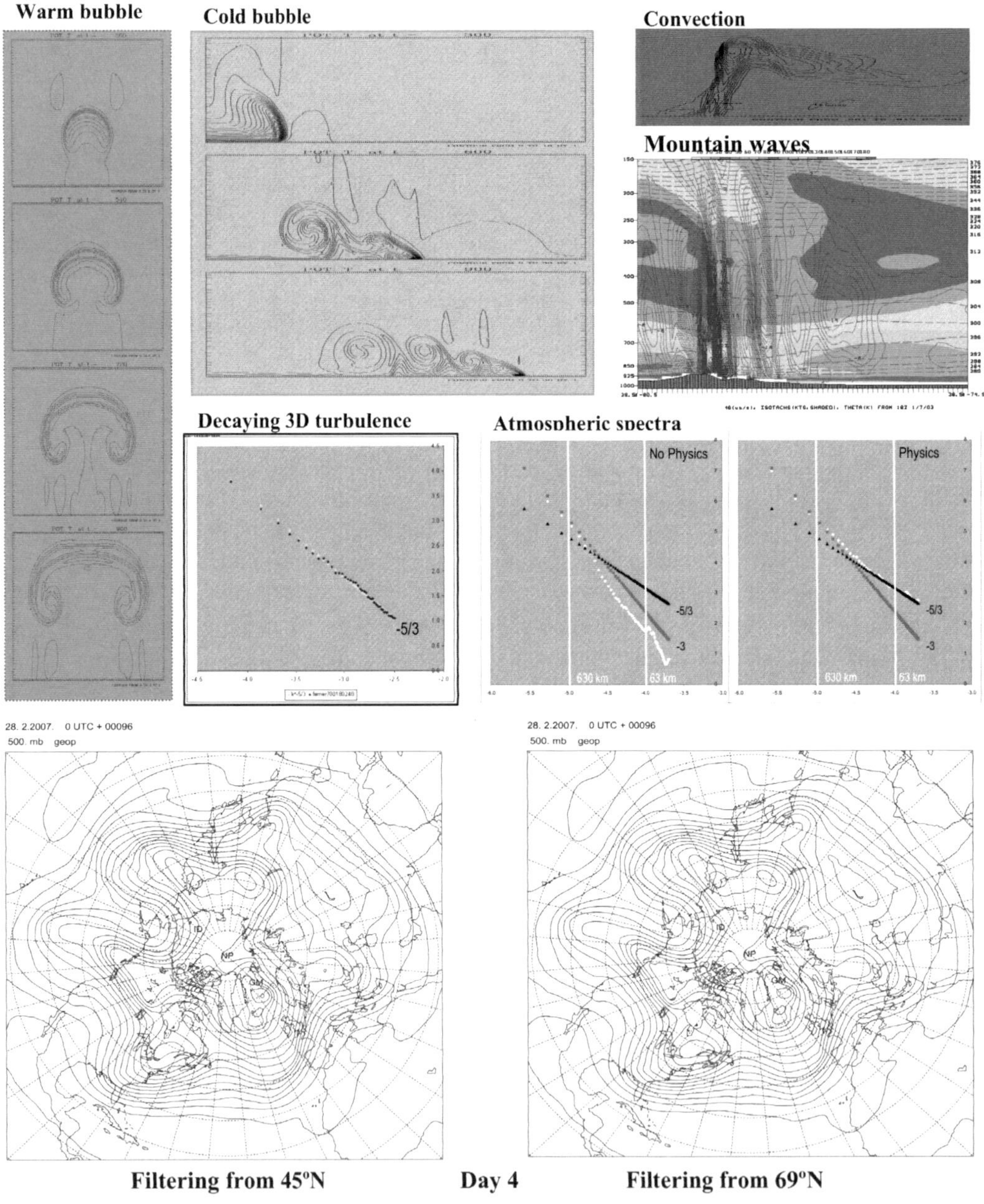

Figure 3. Illustrations of performance of the model formulation on various scales, from classical 2D nonhydrostatic solutions and convective systems, to regional and global forecasts and simulations.

The model code has been parallelized and it scales well. Estimates indicate that the model can be competitive in computational efficiency with other global and regional models. The results of the preliminary testing and evaluation of the model have been encouraging. Some results illustrating the performance of the model formulation on various scales, from classical 2D nonhydrostatic solutions and convective systems, to regional and global forecasts and simulations, are shown in figure 3. The high computational efficiency of the model promises the possibility of application of nonhydrostatic dynamics on the global scale when single digit resolutions become affordable.

3. Some outstanding modeling problems related to atmospheric transport of aerosols

The advective transport of passive scalars received a significant attention of applied mathematicians and atmospheric modelers, and a variety of numerical algorithms have been proposed (see e.g. [13,14,15,16,17,18,19,20,21]). However, fully satisfactory algorithms for all situations and flow regimes have not yet been found and universally accepted.

The transport of "passive" scalars should be conservative, positive definite, monotone, sufficiently accurate and computationally efficient. The Lagrangian approach can easily satisfy the requirements for positive definitness and monotonicity, but difficulties arise with strict conservation and the treatment of open boundary conditions. On the other hand, the Eulerian approach has problems with preservation of positive definitness and monotonicity.

The problems are not restricted only to numerical methods. High resolution mesoscale runs using single digit grid sizes that are increasingly becoming affordable reveal problems with the treatment of moist convection. Namely, the conventional parameterizations of moist convection are inadequate for such scales and the models are struggling to reproduce the convection explicitly, without a parameterized convection (e.g., [22]).

Another physics problem is how to treat the contribution of the coherent structures in the vertical transports in the PBL. At the resolutions approaching the scales of the coherent structures, these structures are becoming partially resolved by the model dynamics and therefore a problem of double counting may arise.

References

[1] Ginoux, P., Chin, M., Tegen, I., Prospero, J., Holben, B., Dubovik, O., and Lin, S.J., 2001 Sources and distributions of dust aerosols simulated with the GOCART model. *Journal. Of Geophyical. Research,* **106,** 20 255–20273.

[2] Janjic, Z., 2005 A unified model approach from meso to global scales. Geophysical Research Abstracts, Vol. 7, 05582, 2005, SRef-ID: 1607-7962/gra/EGU05-A-05582.

[3] Janjic, Z. and Black, T., 2007 An ESMF unified model for a broad range of spatial and temporal scales. *Geophysical Research Abstracts*, Vol. **9**, 05025, 2007, SRef-ID: 1607-7962/gra/EGU2007-A-05025.

[4] Nickovic, S., Kallos, G., Papadopoulos, A., and Kakaliagou, O., 2001 A model for prediction of desert dust cycle in the atmosphere. *Journal of Geophysical Research,* **106**,18113–18129.

[5] Janjic, Z.I., 1979 Forward-backward scheme modified to prevent two-grid-interval noise and its application in sigma coordinate models. *Contrib. Atmos. Phys*., **52**, 69-84.

[6] Janjic, Z.I., 1984 Non-linear advection schemes and energy cascade on semi-staggered grids. *Mon. Wea. Rev*., **112**, 1234-1245.

[7] Mesinger, F., Z.I. Janjic, S. Nickovic, D. Gavrilov and D.G. Deaven, 1988 The step-mountain coordinate: model description and performance for cases of Alpine lee cyclogenesis and for a case of an Appalachian redevelopment. *Mon. Wea. Rev.,* **116**, 1493-1518.

[8] Janjic, Z.I., 1990 The step-mountain coordinate: physical package. *Mon. Wea. Rev.,* **118**, 1429-1443.

[9] Janjic, Z.I., 1994 The step-mountain eta coordinate model: further developments of the convection, viscous sublayer and turbulence closure schemes. *Mon. Wea. Rev*., **122**, 927-945.

[10] Janjic, Z.I., 2001 Nonsingular Implementation of the Mellor-Yamada Level 2.5 Scheme in the NCEP Meso model. NOAA/NWS/NCEP Office Note 437, 61 pp.
[11] Janjic, Z.I., Gerrity Jr., J.P., and Nickovic, S., 2001 An Alternative Approach to Nonhydrostatic Modeling. *Mon. Wea. Rev.*, **129**, 1164-1178.
[12] Janjic, Z.I., 2003 A Nonhydrostatic Model Based on a New Approach. *Meteorology and Atmospheric Physics,* **82**, 271-285. http://dx.doi.org/10.1007/s00703-001-0587-6
[13] Boris, J. P., and Book, D.L., 1973 Flux corrected transport I, SHASTA, a fluid transport algorithm that works. *J. Comput. Phys*., **11**, 38-69.
[14] Bott, A., 1993 The Monotone Area-preserving Flux-Form Advection Algorithm: Reducing the Time-splitting Error in Two-Dimensional Flow Fields. *Mon. Wea. Rev.,* **121**, 2637–2641.
[15] Schneider, H. R., 1984 A Numerical Transport Scheme which Avoids Negative Mixing Ratios. *Mon. Wea. Rev.,* **112**, 1206–1217.
[16] Lin, S-J, and Rood, R.B., 1996 Multidimensional Flux-Form Semi-Lagrangian Transport Schemes. *Mon. Wea. Rev.*, **124**, 2046–2070.
[17] Colella, P., 1990 Multidimensional upwind methods for hyperbolic conservation laws. *J. Comput. Phys.,* **54**, 174-201.
[18] Smolarkiewicz, P. K., and Grabowski, W.W., 1990 The multidimensional positive definite advection transport algorithm: Nonoscillatory option. *J. Comput. Phys*., **86**, 335-375.
[19] Staniforth, A. and Côté, J., 1991 Semi-Lagrangian Integration Schemes for Atmospheric Models—A Review. *Mon. Wea. Rev.,* **119**, 2206–2223.
[20] Van Leer, B., 1977 Toward the ultimate conservative difference scheme. Part IV: A new approach to numerical convection. *J. Comput. Phys*., **23**, 276-299.
[21] Zalesak, S. T., 1979 Fully multidimensional flux-corrected transport algorithms for fluid. *J. Comput. Phys*., 31, 335-362.
[22] Bryan, G. H., Wyngaard, J.C., and Fritsch, J.M., 2003 Resolution requirements for the simulation of deep moist convection. *Mon. Wea. Rev*., **131**, 2394–2416.

IOP Conf. Series: Earth and Environmental Science 7 (2009) 012012 doi:10.1088/1755-1307/7/1/012012

Ten-year operational dust forecasting - Recent model development and future plans

G Kallos, C Spyrou, M Astitha, C Mitsakou, S Solomos, J Kushta, I Pytharoulis, P Katsafados, E Mavromatidis, N Papantoniou and G Vlastou

University of Athens, School of Physics, Atmospheric Modeling and Weather Forecasting Group – UOA/AM&WFG, University Campus, Bldg. PHYS-V, Athens 15784, Greece

E-mail: kallos@mg.uoa.gr

Abstract. The Sahara desert is one of the major sources of mineral dust on Earth, producing up to $2x10^8$ t yr^{-1}. A combined effort has been devoted during the last ten years at the University of Athens (UOA) from the Atmospheric Modeling and Weather Forecasting Group (AM&WFG) to the development of an analysis and forecasting tool that will provide early warning of Saharan dust outbreaks. The developed tool is the SKIRON limited-area forecasting system, based on the Eta limited area modeling system with embedded algorithms describing the dust cycle. A new version of the model is currently available, with extra features like eight-size particle bins, radiative transfer corrections, new dust source identification and utilization of rocky soil characterization and incorporation of more accurate deposition schemes. The new version of SKIRON modeling system is coupled with the photochemical model CAMx in order to study processes like the shading effect of dust particles on photochemical processes and the production of second and third generation of aerosols. Moreover, another new development in the AM&WFG is based on the RAMS model, with the incorporation of processes like dust and sea-salt production, gas and aqueous phase chemistry and particle formation. In this study, the major characteristics of the developed (and under development) modeling systems are presented, as well as the spatiotemporal distribution of the transported dust amounts, the interaction with anthropogenically-produced particles and the potential implications on radiative transfer.

1. Ten-Year Operational Dust Forecasting

Saharan desert is known as the most important source of soil particles in the atmosphere. Saharan dust particles have been associated with changes in the earth's radiation balance, more frequent precipitation events at the affected areas, modifications in photolysis rates, air quality degradation and reduction of visibility. Large concentrations of mineral dust can be found in the tropical Atlantic, reaching the Caribbean due to specific wind patterns. Episodic transport of dust particles to the Mediterranean region, as well as northern Europe has been also observed.

The production, transport and deposition cycle of Saharan dust is rather complicated and the complexity of these processes becomes more enhanced by taking into account their mixture with anthropogenically-produced air pollutants. Advanced modeling techniques, as complementary tools to observations, are implemented world-wide for the identification of production, transport, transformation and deposition patterns, as well as for the understanding of the aerosol cycle in the

© 2009 IOP Publishing Ltd

atmosphere and its potential impacts and interactions with other pollutants. At the framework of well known research projects namely SKIRON, MEDUSE and ADIOS (1995 – 2003), a combined effort was devoted at the UOA/AM&WFG to the development of an analysis and forecasting tool that will provide early warning of Saharan dust outbreaks. The developed tool is the SKIRON limited-area forecasting system. It is based on the Eta limited area modeling system [1, 2 and references therein] with embedded algorithms describing the dust cycle. The prototype system was set in operational use on 1997. At this version, only one particle size was used in the hydrostatic, non-parallelized version of the Eta model. On 2000, a more sophisticated system replaced the older one, developed on parallelized version of the atmospheric model, with new schemes concerning dust particle production in four particle size bins. It is in use in several other Universities and Institutes with the name DREAM from the name of the dust module and the terminology used in Nickovic et al. [3]. This model was continuously improved during the last seven years and new features have been added (e.g. nudging of high resolution Sea Surface Temperature, spatial distributions of clay amounts, in-cloud deposition efficiency). On 2002, a non-hydrostatic version of the SKIRON system was available. This version was in operational use until September 2007 over a large domain, covering North Africa, Mediterranean and Europe with forecasting horizon of three days. The model outputs are widely used by several oceanographic, air quality and ecosystems management Institutes and Agencies of all type. The web page http://forecast.uoa.gr has 6000 – 10000 visitors per day, while the model outputs are available through an FTP server. Monthly and annual maps and statistics have been prepared during the last 7 years and have been made available to UNEP (United Nations Environmental Programme) after a request. Since September 2007 the newer version of the SKIRON model runs operationally, including newly developed features that are discussed in the next section.

In order to derive the seasonal variability of Saharan dust deposition by implementing SKIRON modeling system on the area of the Mediterranean basin, the mean monthly deposition (dry and wet) rates were estimated (figure 1a, b). The basin was separated into four zones, namely West, Central/South, Central/North, East. Dust deposition depicts a clear maximization during the transient spring period and a secondary maximum in the autumn months. During these time periods, due to differential heating between the land of Northern Africa, Southern Europe and the Mediterranean a southern flow is established that enhances the transported amounts of dust towards the Mediterranean sea. According to the four-size bin classification implemented for the years 2003 – 2006, the lighter particles appear to be more likely transported towards the Mediterranean region and Europe and be deposited there.

The amounts of dry and wet deposition of dust over the Mediterranean sea and Europe have been estimated based on the SKIRON system simulations, for a seven-year period (2000 – 2006) (table 1).

Table 1. Annual dust deposition (total, dry and wet) for 2000 – 2006 on the Mediterranean Sea and Europe (in 10^3 t).

	Mediterranean Sea			Europe		
	total	dry	wet	total	dry	wet
2000	3962	1541	2421	3914	936	2978
2001	2500	1255	1245	2909	725	2184
2002	5132	1546	3586	4723	883	3840
2003	6752	1828	4924	1999	256	1743
2004	9306	1504	7802	4933	398	4535
2005	6833	1601	5232	3937	355	3582
2006	6361	1634	4727	2849	256	2593

The analysis showed that the dust amounts deposited over the Mediterranean waters are of the order of 10^7 t yr^{-1}. Similar amounts of dust particles are crossing the Mediterranean region and are being deposited over Europe. Guerzoni et al. [4] have reported noticeably higher values for the dust deposited mass on the Mediterranean region. The differences in the dust calculated amounts are due to the higher spatial resolution and the incorporation of the four-size classification of particle diameters at the present model that can offer more accurate calculations. The prognostic capabilities of the modeling system have been evaluated via the simulations of long-range dust transport in the Mediterranean region [5] and the North-Atlantic ocean [6].

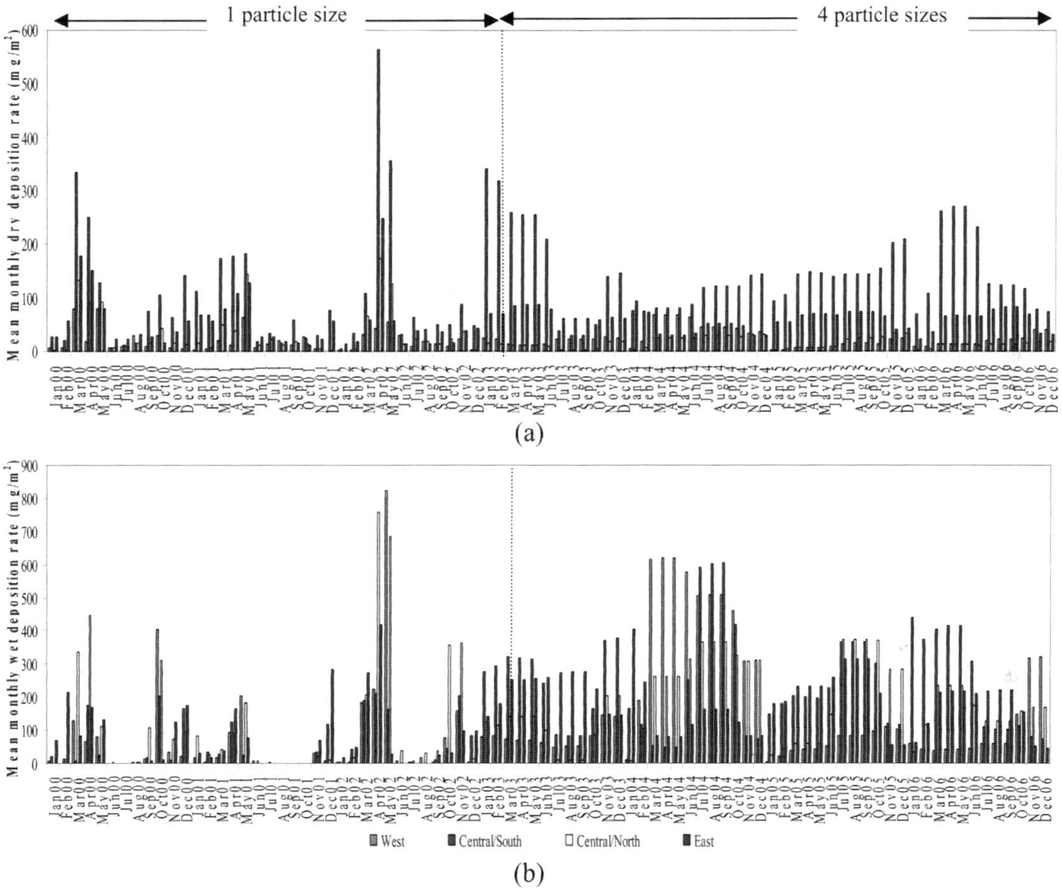

Figure 1. Mean monthly deposition (a) dry and b) wet) for 2000 – 2006 for West, Central/South, Central/North, East Mediterranean basin.

2. New Model Development and Future Plans

Continuous development of the entire SKIRON modeling system has led to a newer model version where inaccuracies existing in previous versions (e.g. [3]) have been corrected and extra features have been added. Some of these include the incorporation of eight size bins of dust particles [7], the calculation of Aerosol Optical Depth (AOD) and the correction of radiative transfer by utilizing look-up tables [8]. Also a new method of dust source identification has been incorporated as well as the utilization of rocky soil characterization, the replacement of the dry and wet deposition schemes with more accurate ones, the in-cloud scavenging, etc. The new version of SKIRON modeling system has been used recently in operational forecasting over a large area from the Caspian Sea to the Western USA and from the tropics to the Arctic Region. The forecasting horizon is five days.

The presence of dust particles in the atmosphere has several effects on radiative transfer processes. Aerosol particles absorb and scatter incoming solar radiation, thus, the radiant energy reaching the surface can be reduced [8]. In the SKIRON model, the Aerosol Optical Depth (AOD) is determined via the particle physical characteristics (radius, density) and the extinction coefficient. In that way, the decrease of the incoming solar radiation is calculated using lookup tables for the different AOD values [8]. The reduction varies as a function of suspended particle concentration, reaching values close to 70 W m^{-2} [9]. In areas with increased dust concentration, the incoming solar flux appears to decrease by up to 40 W m^{-2} (figure 2a). This, in return, reduces the dust flux in the source areas and inhibits dust aerosol production (figure 2b).

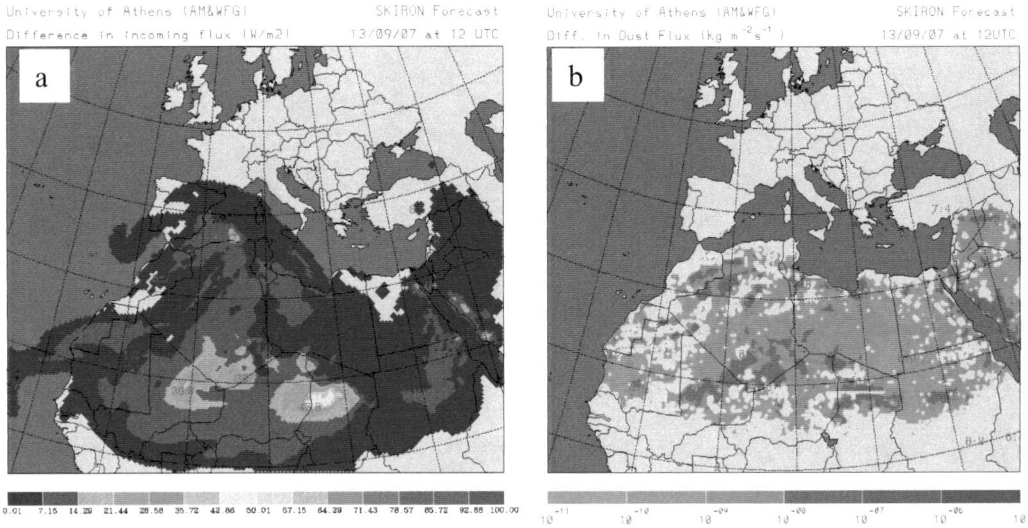

Figure 2. Differences in (a) the incoming solar flux at the surface due to the presence of desert dust particles in W m^{-2} and (b) the dust production due to the decrease of the amount of incoming solar radiation in kg m^{-2} s^{-1}.

The coexistence of natural PM like Saharan dust and anthropogenic pollutants like ozone and sulfates can cause significant direct and indirect impacts on air quality and the environment. Saharan dust can influence physicochemical processes on various ways. One is the shading effect that alters the radiation and causes modifications of the photolysis rates. In that way, desert dust influences the production of species like ozone, nitrogen oxides and particulate sulfate (figure 3a-b). A more complicated impact of increased dust load in the atmosphere is through the heterogeneous chemical reactions that occur in the surface of a wet dust particle. Such heterogeneous reactions lead to the formation of new types of aerosols, for example dust particle coated with sulfate (figure 3c). The SKIRON modeling system is coupled with the photochemical model CAMx [10] in order to study the above mentioned processes [11,12,13,14].

In addition to the improvements in the SKIRON modeling system, a major model development is in progress at the framework of CIRCE project. This development has RAMS model as atmospheric driver where the dust cycle modules from SKIRON are directly coupled together with sea-salt, gas and aqueous phase chemistry and particle formation. For the testing of the above modules, several test runs are performed. A severe dust transport episode has occurred on April 16-18, 2005 towards Eastern Mediterranean and Greece. Measurements of PM10 in several stations near the Athens International Airport (AIA) ranged from 1000 to 3000 μg m^{-3} during April 17, 2005. This case has been simulated using the new RAMS model. Figure 4 shows the vertical distribution of modeled PM10 dust concentration for different hours of the day (April 17, 2005), at the AIA. Maximum PM10 dust concentration near the surface has occurred on April 17, 2005, at 12:00 UTC, according to the model results. The new model development will also include Cloud Condensation Nuclei (CCN) and Ice Nuclei (IN) concentrations as prognostic parameters directly coupled with the explicit cloud microphysical scheme existing in RAMS. The new model development will be used for studying feedbacks between air pollution and climate.

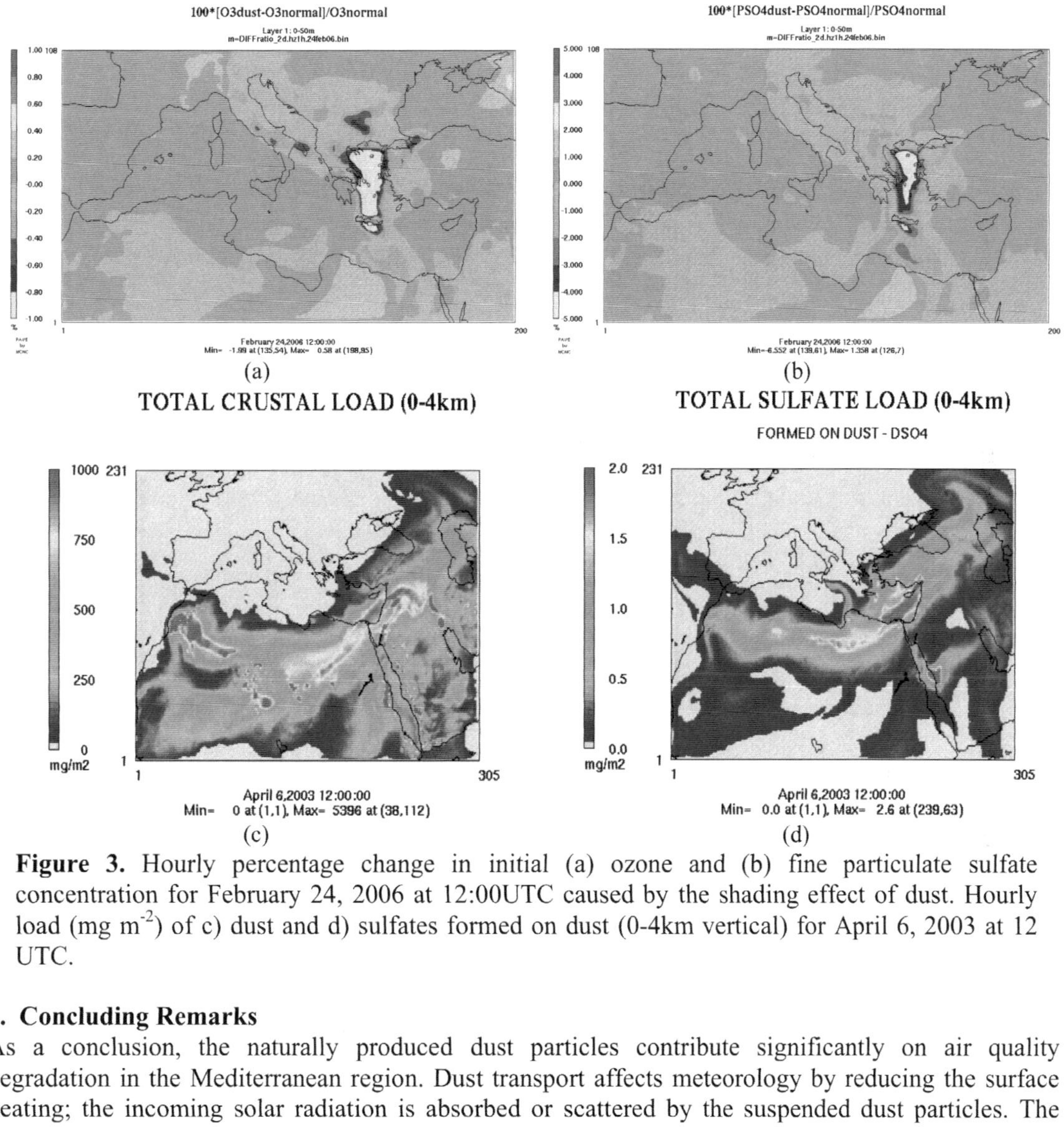

Figure 3. Hourly percentage change in initial (a) ozone and (b) fine particulate sulfate concentration for February 24, 2006 at 12:00UTC caused by the shading effect of dust. Hourly load (mg m^{-2}) of c) dust and d) sulfates formed on dust (0-4km vertical) for April 6, 2003 at 12 UTC.

3. Concluding Remarks

As a conclusion, the naturally produced dust particles contribute significantly on air quality degradation in the Mediterranean region. Dust transport affects meteorology by reducing the surface heating; the incoming solar radiation is absorbed or scattered by the suspended dust particles. The systematic inaccuracies of the previous versions have been corrected in the new version of the SKIRON modeling system and new features have been included. The new RAMS development is expected to provide a tool that will have some unique features to study desert dust and impacts on clouds and radiation that do not exist in current modeling systems.

Acknowledgements

This work is funded by European Union under the project CIRCE (EC contract no. 036961) and the Hellenic Ministry of Development, General Secretariat for Research and Technology under the project PENED 2003 – Operational Program Competitiveness. This effort is also supported by EUROCONTROL.

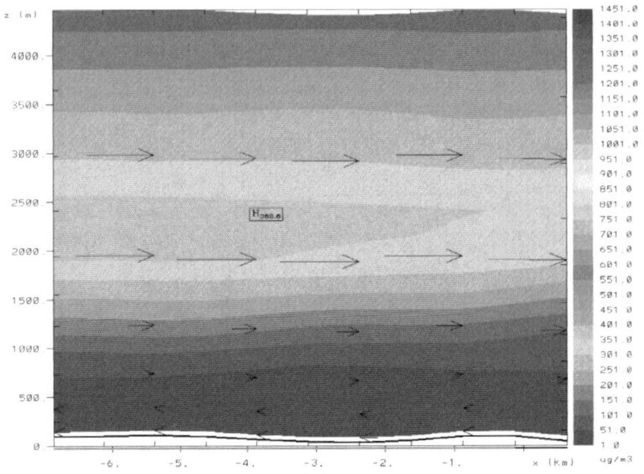

(a)

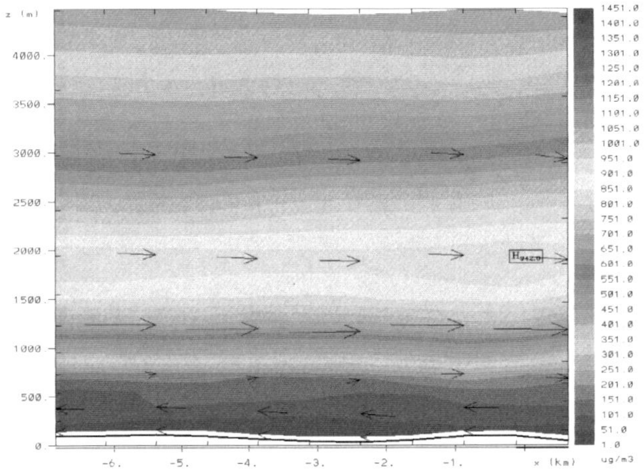

(b)

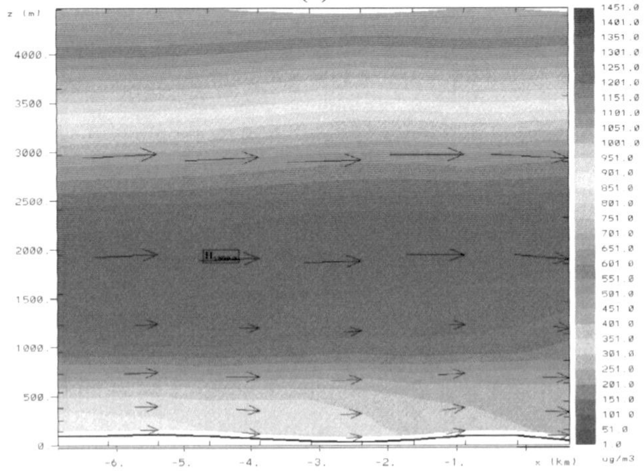

(c)

Figure 4. Vertical cross sections of PM10 concentrations at the Athens International Airport on April 17, 2005 at (a) 00:00UTC, (b) 06:00UTC and (c) 12:00UTC.

References

[1] Mesinger, F., Janjic, Z.I., Nickovic, S., Gavrilov, D., and Deaven, D.G., 1988 The step-mountain coordinate: Model description and performance for cases of Alpine lee cyclogenesis and for a case of an Appalachian redevelopment. *Mon. Weather Rev.*, **116**, 1493-1518.

[2] Janjic, Z., 1994 The step-mountain eta coordinate model: Further developments of the convection, viscous sublayer and turbulence closure schemes. *Mon. Weather Rev.*, **122**, 927-945.

[3] Nickovic, S., Kallos, G., Papadopoulos, A., and Kakaliagou, O., 2001 A model for prediction of desert dust cycle in the atmosphere. *J. Geophys. Res.*, **106**,18113–18129.

[4] Guerzoni, S., R. Chester, F. Dulac, B. Herut, M.-D. Loye-Pilot, C. Measures, C. Migon, E. Molinaroli, C. Moulin, P. Rossini, C. Saydam, A. Soudine, and P. Ziveri, 1999. The role of atmosphere deposition in the biogeochemistry of the Mediterranean Sea. *Prog. Oceanogr.* **44**, 147-190.

[5] Kishcha P., P. Alpert, A. Shtivelman, S.O. Krichak, J.H. Joseph, G. Kallos, P. Katsafados, C. Spyrou, G.P. Gobbi, F. Barnaba, S. Nickovic, C. Perez, and J.M. Baldasano, 2007 Forecast errors in dust vertical distributions over Rome (Italy): Multiple particle size representation and cloud contributions. *J. Geophys. Res.*, **112**, D15205, doi: 10.1029/2006JD007427.

[6] Kallos, G., A. Papadopoulos, P. Katsafados and S. Nickovic, 2005 Trans-Atlantic Saharan dust transport: Model simulation and results. *J. Geophys. Res.*, **111**, D09204, doi: 10.1029/2005JD006207.

[7] Zender, C. S., H. Bian, and D. Newman, 2003 Mineral Dust Entrainment and Deposition (DEAD) model: Description and 1990s dust climatology. *J. Geophys. Res.*, **108**, D14, doi:10.1029/2002JD002775.

[8] Kaufman, Y. J., D. Tanre, B.N. Holben, S.Mattoo, L. A. Remer, T. F. Eck, J. Vaughan and Bernadette Chatenet, 2002 Aerosol radiative impact on spectral solar flux at the surface, derived from principal-plane sky measurements. *J. Atmos. Sci.*, **59**, 635-646.

[9] Tegen, I., A.A. Lacis, and I. Fung, 1996 The influence of mineral aerosols from disturbed soils on the global radiation budget. *Nature*, **380**, 419-422, doi: 10.1038/380419a0.

[10] Environ 2006: User's Guide to the Comprehensive Air Quality Model with Extensions (CAMx). Version 4.31. Prepared by ENVIRON International Corporation, Novato, CA.

[11] Astitha M., G. Kallos, P. Katsafados, I. Pytharoulis and N. Mihalopoulos, 2006 Radiative effects of natural PM on photochemical processes in the Mediterranean Region. 28th NATO/CCMS ITM, May 2006, Leipzig, Germany.

[12] Astitha M., G. Kallos, P. Katsafados, and E. Mavromatidis, 2007 Heterogeneous chemical processes and their role on particulate matter formation in the Mediterranean Region. 29th NATO/CCMS ITM September 2007, Aveiro, Portugal.

[13] Astitha M. and G. Kallos, 2008: Gas-phase and aerosol chemistry interactions in South Europe and the Mediterranean Region. *Environmental Fluid Mechanics*, DOI: 10.1007/s10652-008-9110-7.

[14] Kallos, G., M. Astitha, P. Katsafados and C. Spyrou, 2007 Long-range transport of anthropogenically and naturally produced particulate matter in the Mediterranean and North Atlantic: Current state of knowledge. *J. Appl. Meteorol. Clim.*, **46**, doi: 10.1175/JAM2530.1.

Dust Modelling and Forecasting in the Barcelona Supercomputing Center: Activities and Developments

C Pérez[1], J M Baldasano[1,2], P Jiménez-Guerrero[1], O Jorba[1], K Haustein[1], E Cuevas[3], S Basart[1] and S Nickovic[4]

[1]Earth Sciences Department. Barcelona Supercomputing Center. Barcelona, Spain.
[2]Laboratory of Environmental Modeling. Universitat Politècnica de Catalunya. Barcelona. Spain
[3]Izaña Atmospheric Research Center. Agencia Estatal de Meteorología. Tenerife, Spain.
[4]Atmospheric Research and Environment Branch. World Meteorological Organization. Geneva, Switzerland.

E-mail: carlos.perez@bsc.es

Abstract. The Barcelona Supercomputing Center (BSC) is the National Supercomputer Facility in Spain, hosting MareNostrum, one of the most powerful Supercomputers in Europe. The Earth Sciences Department of BSC operates daily regional dust and air quality forecasts and conducts intensive modelling research for short-term operational prediction. This contribution summarizes the latest developments and current activities in the field of sand and dust storm modelling and forecasting.

1. Dust modelling and forecasting at BSC

The Earth Sciences Department of BSC maintains dust forecast operations with the Dust REgional Atmospheric Model (DREAM) and conducts modelling research and developments for short-term prediction.

BSC-DREAM [1] predicts the atmospheric life cycle of the eroded desert dust and was developed as a pluggable component of the Eta/NCEP (National Centers for Environmental Prediction) model. BSC-DREAM solves the Euler-type partial differential non-linear equation for dust mass continuity and it is fully inserted as one of the governing prognosis equations in the atmospheric Eta/NCEP atmospheric model equations [2,3,4,5,6,7,8,9,10].

BSC-DREAM simulates all the major processes of the atmospheric dust cycle. During model integration, calculation of the surface dust injection fluxes is made over the model grid points declared as deserts. Once injected into the air, dust aerosol is driven by the atmospheric model variables: by turbulent parameters in the early stage of the process when dust is lifted from the ground to the upper levels; by model winds in the later phases of the process when dust travels away from the sources; finally, by thermodynamic processes and rainfall of the atmospheric model and land cover features which provide wet and dry deposition of dust over the Earth surface.

One of the key components of the dust model is the treatment of the sourcing terms in the concentration continuity equation. Failure to adequately simulate/predict the production phase of the dust cycle leads to wrong representation of all other dust processes in the model. Therefore, special

© 2009 IOP Publishing Ltd

attention is made to properly parameterize dust production phase. Wind erosion of the soil in BSC-DREAM parameterization scheme is controlled mainly by the following factors: type of soil, type of vegetation cover, soil moisture content, and surface atmospheric turbulence. In the model, grid points acting as desert dust sources are specified using arid and semiarid categories of the global USGS 1-km vegetation data set. Another data participating in dust production calculations is the FAO 4-km global soil texture data set from which particle size parameters are evaluated. The main general features of the model are listed below:

- Dust production scheme with introduced viscous sub-layer [11,6].
- 8 particle size bin distribution.
- Soil wetness effects on dust production [12].
- Dry deposition [13] and below cloud scavenging.
- Horizontal and vertical advection, turbulent and lateral diffusion [6] represented as for other scalars in the Eta/NCEP model.
- Dust radiative feedbacks on meteorology [14].

Eight size transport bins between 0.1 and 10 μm are considered. In this interval, the aerosol effects on solar radiation are the most significant. Within each transport bin, dust is assumed to have time-invariant, sub-bin log-normal distribution [15] employing the transport mode with mass median diameter of 2.524 μm [16] and geometric standard deviation 2.0 [17]. The analytic sub-bin distribution allows accurate prescription of physical and optical properties known to vary across the bin width.

Since there are not yet satisfactory three-dimensional dust concentration observations, the initial state of dust concentration in the model is defined by the 24-hour forecast from the previous-day model run. The model at the starting day is run using "cold start" conditions, i.e., the zero-concentration initial state. The model needs about 2-3 days for spinning up and for establishing reliable dust concentration conditions.

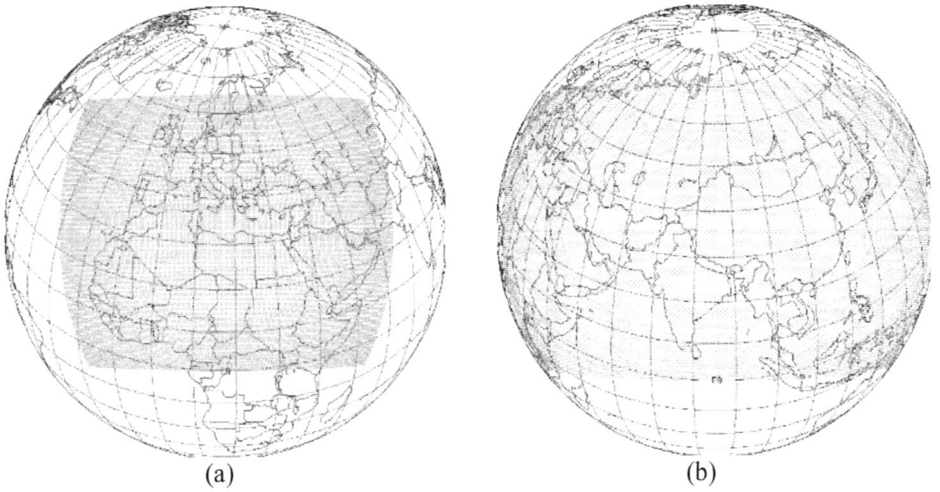

(a) (b)

Figure 1. Northern Africa-Europe-Mediterranean-Middle East (a) and East-Asia (b) forecast domains of BSC-DREAM.

BSC-DREAM has been delivering operational dust 72h forecasts over North Africa-Europe-Mediterranean (which include North of Africa, the Mediterranean basin, Middle East and Europe) and Asia regions (figure 1) in the last years (http://www.bsc.es/projects/earthscience/DREAM). The operational resolution is 0.3°x0.3° and 24 layers up to 15 km in the vertical.

IOP Conf. Series: Earth and Environmental Science 7 (2009) 012013 doi:10.1088/1755-1307/7/1/012013

The model qualitative and quantitative verification studies performed so far [18,19,20] using data from observation networks such as the European Lidar Network EARLINET, the AERONET/PHOTONS sun photometer network, satellite and ground level PM levels (EMEP) outline the good skills of the model concerning both the horizontal and vertical extent of the dust plume in the geographic region of application. Additionally, the model has been validated and tested against measurements at source regions for SAMUM [21] and BODEX campaigns [22].

In order to improve of the dust forecast and implement new model versions and operational products, daily verification with near-real time observations (figure 2) is conducted in collaboration with the Spanish Meteorological State Agency (AEMET) and the Spanish National Research Council (CSIC). Currently, the verification system includes satellites (MODIS, MSG, OMI) and sun photometers (AERONET). Ongoing activities involve the inclusion of lidars, visibility reports and surface concentration measurements to the verification system.

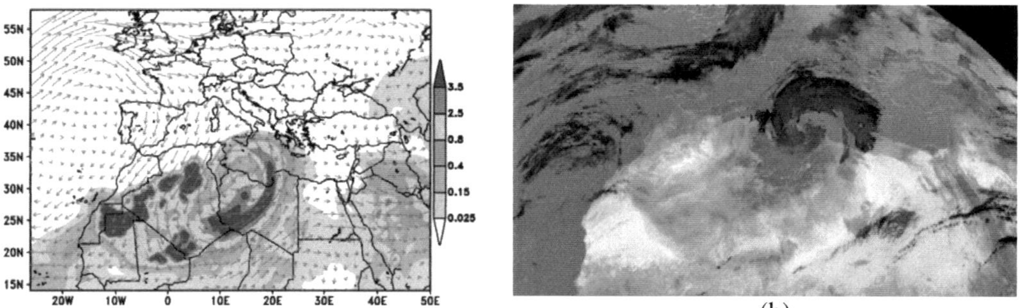

(a) (b)

Figure 2. 24h BSC-DREAM forecast (a) and MSG dust RGB product (b) for 9th March 2007 at 12 UTC.

2. Inclusion of Saharan dust in an integrated air quality forecasting system for Europe

One of the most recent development activities related to dust modelling has focused on the implementation of a high-resolution air quality forecasting system for Europe and the Iberian Peninsula (CALIOPE Project supported by the Spanish Ministry of Environment) [23,24]. The dust contribution to PM levels is included through coupling of the WRF-ARW/HERMES-EMEP/CMAQ/BSC-DREAM modelling system. The high resolution applied (12 km for Europe and 4 km for the Iberian Peninsula) helps providing a better understanding of the dynamics of atmospheric pollutants and allows for early communication to the dwellers in areas that exceed air quality thresholds. This service is available on the website http://www.bsc.es/caliope.

The complex atmospheric conditions in southern Europe involve the superposition of phenomena occurring at different scales that exert a high influence in the levels of particulate matter and its composition. Namely, the contribution of mineral aerosols is very high due to the existence of semiarid soils and specially the frequent occurrence of Saharan dust events in southern Europe. When considering only anthropogenic emissions, chemistry-transport model simulations recursively underestimate the PM10 and PM2.5 concentrations using the current knowledge about aerosol physics and chemistry. Therefore, the introduction of boundary conditions for Saharan dust is necessary in order to model correctly the PM mass over Europe. Dust peaks cannot be represented by introducing boundary conditions derived from a dust climatology due to the highly episodic nature of the events in the region (1-4 days average duration).

The necessity of coupling both aforementioned systems in an integrated framework was addressed by the study of year 2004 [20]. Figure 3 shows the average PM10 concentration in Europe with and without dust contribution.

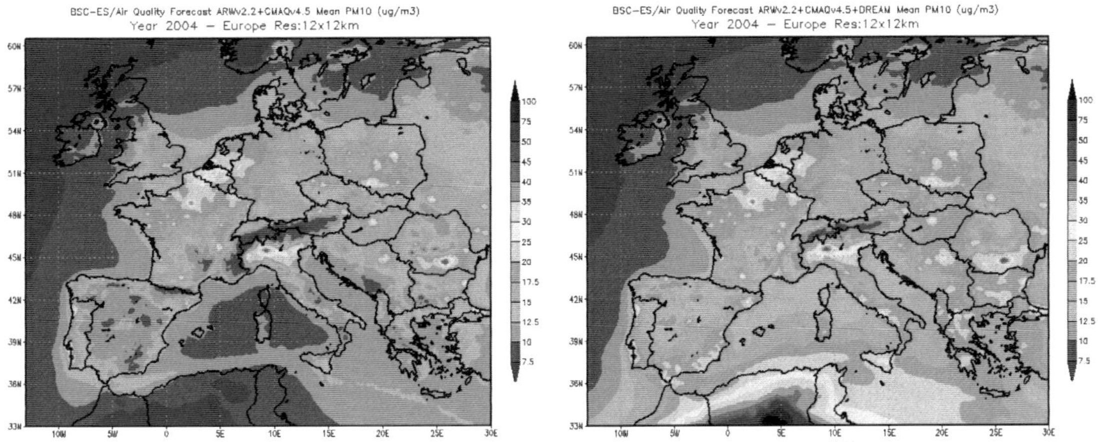

(a) (b)

Figure 3. 2004 average PM10 Concentration (μg m^{-3}) in Europe: anthropogenic contribution with CMAQ (a) and anthropogenic+dust levels with CMAQ+BSC-DREAM (b).

The performance of the model has been quantitatively evaluated with discrete and categorical (skill scores) statistics by a comparison of the first-layer simulations results of CMAQ+BSC-DREAM and the values measured in the European stations belonging to the EMEP network with available data for the year 2004 (26 stations for PM10 and 17 stations for PM2.5). The results indicate an improvement in the discrete statistics for particulate matter (the normalised bias improves from -20% to -9% and the normalised error from 36% to 30% when considering all the stations in Europe for an annual cycle) and skill-scores evaluation (accuracy, critical success index and probability of detection) of PM10 and PM2.5 concentrations when using CMAQ+BSC-DREAM compared to CMAQ-alone simulations. This improvement in the quality of the predictions becomes more important in those southern stations undergoing larger impacts of Saharan dust outbreaks (the normalised error improves from 40-60% to 30-40% for PM10 and from 30-40% to 25-30% for PM2.5 concentrations). As an example, figure 4 shows the comparison at Viznar station in southern Spain.

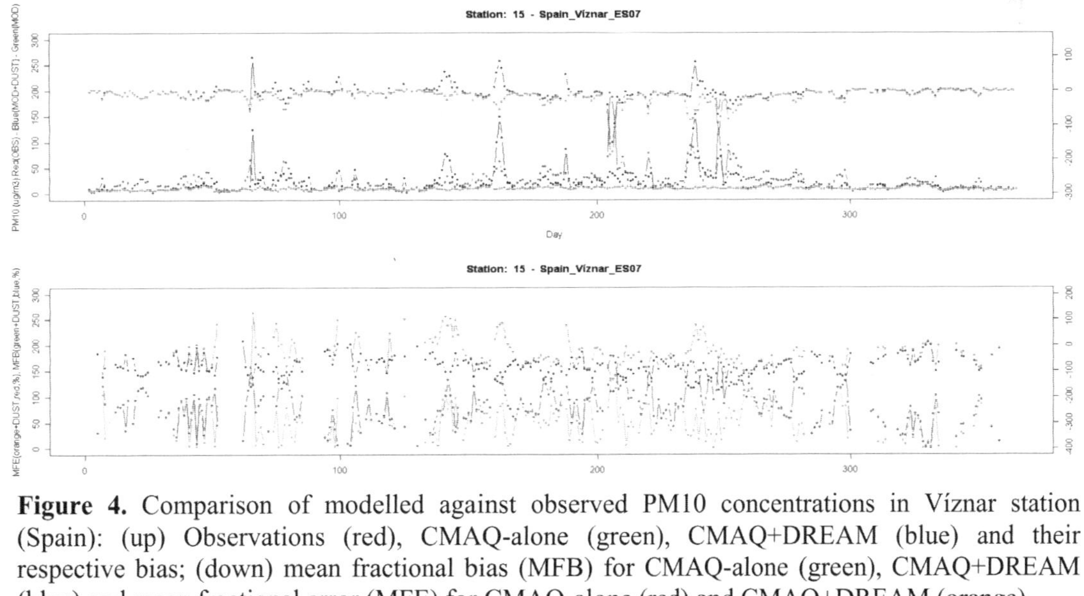

Figure 4. Comparison of modelled against observed PM10 concentrations in Víznar station (Spain): (up) Observations (red), CMAQ-alone (green), CMAQ+DREAM (blue) and their respective bias; (down) mean fractional bias (MFB) for CMAQ-alone (green), CMAQ+DREAM (blue) and mean fractional error (MFE) for CMAQ-alone (red) and CMAQ+DREAM (orange).

The results of this work demonstrated that this first approach to the coupling of CMAQ anthropogenic outputs with BSC-DREAM natural dust forecasts in an operational way is necessary in order to improve the prediction of the particulate matter mass over southern Europe and to achieve the standards set in European Directives for modelling applications.

3. Impact of mineral dust on numerical weather forecasts

Several studies have suggested that the inclusion of mineral dust radiative effects could lead to a significant improvement in the radiative balance of numerical weather prediction models with subsequent improvements in the weather forecast itself. Recently, [25] related systematic short-term temperature forecast errors to the absence of dust radiative effects in models over the Saharan region. Also, [26] stated that the neglect of mineral dust in a NWP model is the most probable reason for the discrepancy in outgoing longwave radiation between the model and Meteosat-7 observations over northern Africa.

In this context, the radiative effects of mineral dust have been fully incorporated into BSC-DREAM. Dust affects the radiative fluxes at the surface and the top of the atmosphere, and the temperature profiles at every time step of the model when the radiation module is processed. These changes influence the atmospheric dynamics, moisture physics and near-surface conditions. Furthermore, dust emission is modified by changes in friction velocity and turbulent exchange coefficients; dust turbulent mixing, transport and deposition are altered by changes in atmospheric stability, precipitation conditions and free-atmosphere winds.

The new scheme was tested for a major dust outbreak over the Mediterranean on April 2002 in order to assess the dust impacts on regional numerical weather forecasting [14]. It is shown that the newly developed atmosphere-dust feedback scheme increases the accuracy of both, atmospheric temperature and mean sea-level pressure forecasts (figure 5). Both low-level warm and upper-level cold temperature biases are considerably reduced when dust affects the atmosphere thermodynamics. The root mean square error of the mean sea level pressure over the whole domain was reduced by almost 20%. The proposed model with integrated dust and atmospheric radiation represents a promising approach for further improvements in numerical weather prediction practice and radiative impact assessment over dust-affected areas.

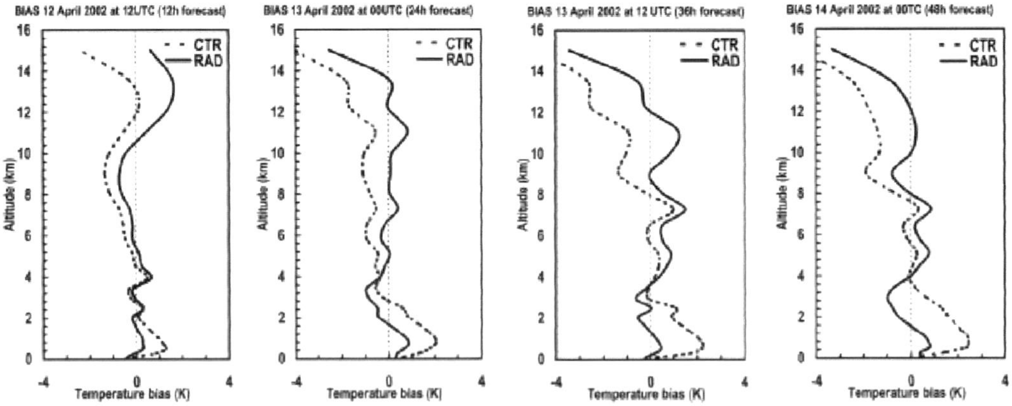

Figure 5. Atmospheric temperature bias of CTR and RAD over the main dust-affected area (comprised between latitude 30°N to 45°N and longitude 0 to 20°E) for the 12, 24, 38, and 48 hour forecasts of the 0000 UTC cycle on 12 April. Extracted from [14].

4. Long term model integrations from 1958 to 2006: validation, analysis and applications

Measurements since the 1960s have shown strong daily, seasonal and inter-annual variations of dust concentration over the Atlantic [27]. [28] used satellite data to outline the correlation between the NAO index and the dust distribution over North Africa and the Atlantic Ocean in the two last decades. [29] found similar results running a dust model for the period 1981-1996.

A 48-year model simulation was performed for the period 1958-2006 at 0.3°x0.3° resolution (implemented in MareNostrum supercomputer) in order to analyze the monthly, seasonal and year-to-year variation of the atmospheric dust load, surface concentration, deposition and the frequency and duration of the events over the Mediterranean and the Eastern North Atlantic [30]. Meteorological fields were initialized every 24 hours and boundary conditions updated every 6 hours with the NCEP/NCAR I global reanalysis (2.5°x2.5° resolution). Satellite observations (figure 6a) and in situ measurements of concentrations at the Izaña station (Canary Islands) for the period 1987-1999 and in southern Europe were used for model validation.

We explored links of the simulated dust parameters to the variability of the large-scale circulation in winter covering decades with mostly positive (1981-2006) and mostly negative (1958-1980) NAO index (figure 6b). The model fairly well represents the interannual winter variability over the Atlantic (figure 6a). In agreement with previous studies we find high NAO dust correlations over the Atlantic Ocean and the southern Sahara/Sahel region within the period 1981-2006 (figure 6c). However, the correlation is significantly decreased when NAO index is mostly negative (1958-1980) (figure 6d).

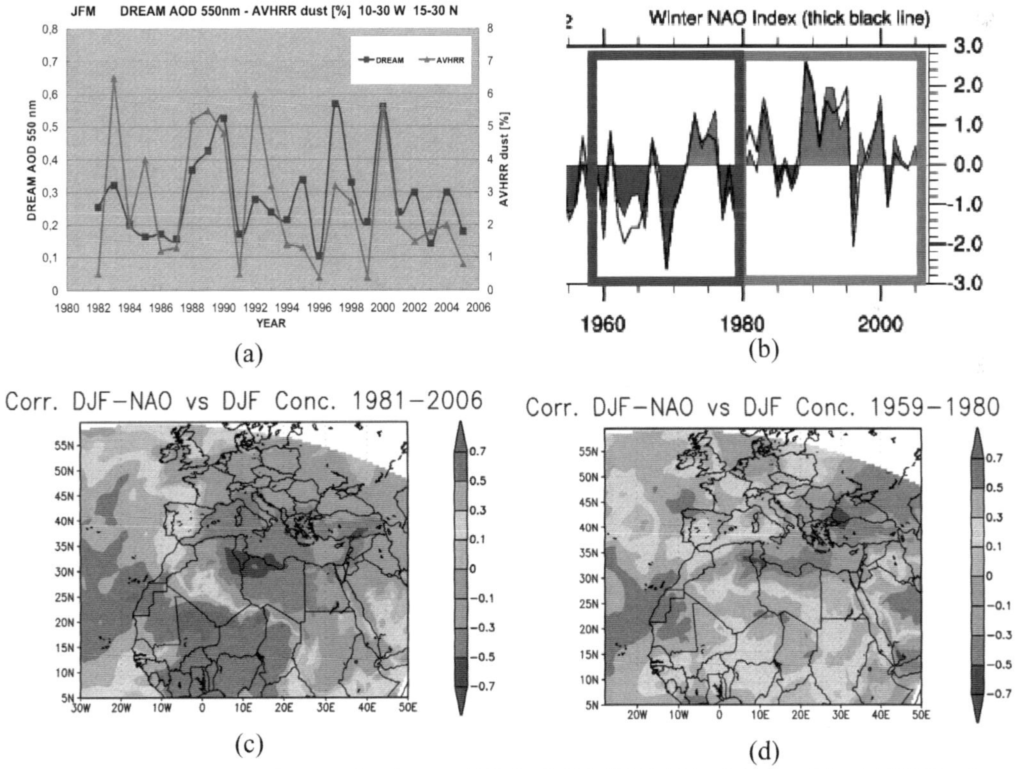

Figure 6. (a) Comparison of winter (JFM) BSC-DREAM AOD with AVHRR dust qualitative product (Evan et al., 2006) over the Eastern Atlantic. Correlation coefficient is 0.62. (b) Winter NAO index. Blue square contains mostly negative phases in the period 1958-1980. Red square contains mostly positive phases in the period 1981-2006. Winter correlation between NAO index and modelled surface dust concentration in the periods 1981-2006 (c) and 1959-1880 (d).

5. Perspectives: an online mineral dust model within the global/regional NMMB

While mineral dust distribution and effects are important on global scales, they strongly depend on dust emissions that are occurring on small spatial and temporal scales. Indeed, the accuracy of surface wind speed used in dust models is crucial. Due to the high-order power dependency on wind friction velocity and the threshold behaviour of dust emissions, small errors in surface wind speed lead to large dust emission errors. Most global dust models use prescribed wind fields provided by major meteorological centres (e.g., NCEP and ECMWF) and their spatial resolution is currently about 1°×1°. Such wind speeds tend to be strongly underestimated over arid and semi-arid areas and do not account for mesoscale systems responsible for a significant fraction of dust emissions regionally and globally. Other significant uncertainties in dust emissions resulting from such approaches are related to the misrepresentation of high subgrid-scale spatial heterogeneity in soil and vegetation boundary conditions, mainly in semi-arid areas.

In order to significantly reduce these uncertainties, BSC is currently implementing a mineral dust model coupled on-line with the new global/regional NMMB (B-grid) atmospheric model using the ESMF framework under development in NOAA/NCEP/EMC [31]. The NMMB is an evolution of the operational WRF-NMME (E-grid) extending from meso to global scales, and including non-hydrostatic option and improved tracer advection. This model is planned to become the next-generation NCEP mesoscale model for operational weather forecasting in North America. Current implementation is based on the well established BSC-DREAM model. First successful global simulations show the potentials of such an approach and compare well to BSC-DREAM regionally (figure 7).

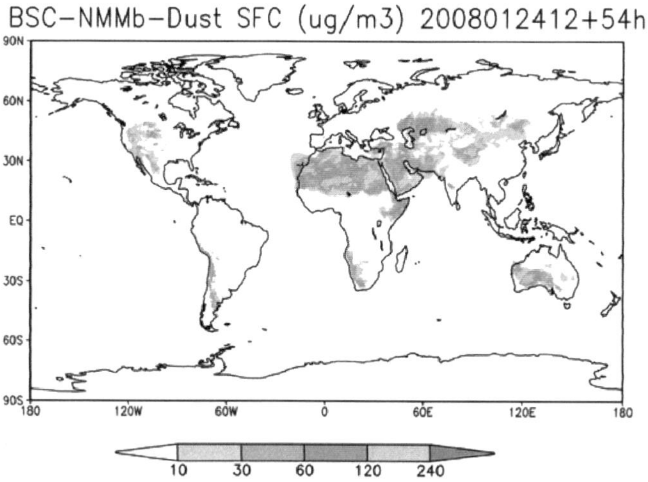

Figure 7. First tests with the NMMb/BSC-Dust model. Surface concentration in μg/m^3 of size bin 4 [0.6–1 μm] for the 54h forecast of the 2008-01-24-12 cycle.

Ongoing developments include improvements in dust size distribution representation, sedimentation, dry deposition, wet scavenging and dust-radiation feedback, as well as the efficient implementation of the model on High Performance Supercomputers for global simulations and forecasts at high resolution.

Acknowledgements

This work was partly funded by the projects CICYT CGL2006-11879 and CGL2006-08903 of the Spanish Ministry of Education and Science and CALIOPE project 441/2006/3-12.1 of the Spanish Ministry of the Environment.

References

[1] Nickovic et al., 2001 A model for prediction of desert dust cycle in the atmosphere. *Journal of Geophysical Research*, **106**, D16, 18113-18129.

[2] Janjic, Z. I., 1977 Pressure gradient force and advection scheme used for forecasting with steep and small scale topography. *Contributions in Atmospheric Physics*, **50**, 186-199.

[3] Janjic, Z. I., 1979 Forward-backward scheme modified to prevent two grid-interval noise and its application in sigma coordinate models. *Contributions in Atmospheric Physics*, **52**, 69-84.

[4] Janjic, Z. I., 1984 Non-linear advection schemes and energy cascade on semi-staggered grids. *Monthly Weather Review*, **112**, 1234-1245.

[5] Janjic, Z. I., 1990 The step-mountain coordinate: Physical package. *Monthly Weather Review*, **118**, 1429-1443.

[6] Janjic, Z. I., 1994 The Step-Mountain Eta Coordinate Model: Further Developments of the Convection, Viscous Sublayer, and Turbulence Closure Schemes. *Monthly Weather Review*, **122**, 927-945.

[7] Janjic, Z. I., 1996 The surface layer parameterization in the NCEP Eta Model. World Meteorological Organisation, Geneva, Switzerland.

[8] Mesinger et al., 1988 The Step-Mountain Coordinate: Model Description and Performance for Cases of Alpine Lee Cyclogenesis and for a Case of an Appalachian Redevelopment. *Monthly Weather Review*, **116**, 1493-1518.

[9] Black, T. L., 1994 The New NMC Mesoscale Eta Model: Description and Forecast Examples. Weather and Forecasting - NMC Notes, 9, 265-278.

[10] Mesinger, E., 1997 Dynamics of limited area models: Formulation and numerical methods. *Journal of Atmospheric Physics*, **63**, 3-14.

[11] Shao, Y., M. R. Raupach and P. A. Findlater, 1993 Effect of saltation bombardment on the entrainment of dust by wind. *Journal of Geophysical Research*, **98**, D7, 12719-12726.

[12] Fecan, F., B. Marticorena and G. Bergametti, 1999 Parameterization of the increase of the aeolian erosion threshold wind friction velocity due to soil moisture for arid and semi-arid areas. *Annales Geophysicae*, **17**, 149-157.

[13] Giorgi, F., A particle dry-deposition parameterization scheme for use in tracer transport models. Journal of Geophysical Research, 91, 9794-9804, 1986.

[14] Pérez, C., S. Nickovic, G. Pejanovic, J. M. Baldasano, and E. Özsoy, 2006 Interactive dust-radiation modeling: A step to improve weather forecasts. *Journal of Geophysical Research*, doi:10.1029/2005JD006717.

[15] Zender, C. S., H. Bian, and D. Newman 2003 Mineral Dust Entrainment and Deposition (DEAD) model: Description and 1990s dust climatology. *Journal of Geophysical Research*, **108**, D14, doi:10.1029/2002JD002775.

[16] Shettle, E. P., 1984 Optical and radiative properties of a desert aerosol model. Proceedings of the Symposium on Radiation in the Atmosphere.

[17] Schulz, M., Y. J. Balkanski, W. Guelle and F. Dulac, 1998 Role of aerosol size distribution and source location in a three-dimensional simulation of a Saharan dust episode tested against satellite-derived optical thickness. *Journal of Geophysical Research*, **103**, D9, 10579-10592.

[18] Pérez, C., S. Nickovic, J. M. Baldasano, M. Sicard, F. Rocadenbosch and V. E. Cachorro. 2006 A long Saharan dust event over the Western Mediterranean: lidar, sun photometer observations and regional dust modeling. *Journal of Geophysical Research*, **111**, D15214, doi:10.1029/2005JD006579.

[19] Balis D., V. Amiridis, S. Kazadzis, A. Papayannis, G. Tsaknakis, S. Tzortzakis, N. Kalivitis, M. Vrekoussis, M. Kanakidou, N. Mihalopoulos, G. Chourdarkis, S. Nickovic, C. Pérez, J.M. Baldasano and M. Drakakis 2006 Optical characteristics of desert dust over the East Mediterranean during summer: a case study. *Annales Geophysicae*, **24**.

[20] Jiménez-Guerrero, P., C. Pérez, O. Jorba and J. M. Baldasano, 2008 Contribution of Saharan dust in an integrated air quality system and its on-line assessment. *Geophysical Research Letters*, **35**(3).

[21] Haustein, K., Pérez, C., Baldasano, J.M. et al., 2009 Regional dust model performance during SAMUM 2006. *Geophysical Research Letters* (in press).

[22] Todd, M. C., D. B. Karam, C. Cavazos, C. Bouet, B. Heinold, J.M. Baldasano, G. Cautenet, I. Koren, C. Perez, F. Solmon, I. Tegen, P. Tulet, R. Washington, and A. Zakey 2008 Quantifying uncertainty in estimates of mineral dust flux: an inter-comparison of model performance over the Bodélé Depression, Northern Chad. *J. Geophys. Res.*, doi:10.1029/2008JD010476, in press.

[23] Baldasano, J.M., Jiménez-Guerrero, P., Jorba, O., Pérez, C., López, E., Güereca, P., Martín, F., García-Vivanco, M., Palomino, I., Querol, X., Pandolfi, M., Sanz, M.J., Diéguez, J.J., 2008 CALIOPE: an operational air quality forecasting system for the Iberian Peninsula, Balearic Islands and Canary Islands. Current status and future plans. *Advances in Science and Research* **2**, 89-98.

[24] Baldasano J.M., L. P. Güereca, E. López, S. Gassó, P. Jimenez-Guerrero 2008 Development of a high resolution (1 km x 1 km, 1 h) emission model for Spain: the High-Elective Resolution Modelling Emission System (HERMES). *Atmospheric Environment*, **42** 7215-7233.

[25] Kishcha et al., 2003 Atmospheric response to Saharan dust deduced from ECMWF reanalysis (ERA) temperature increments. *Tellus*, **55**, Serie B, 901-913.

[26] Haywood et al., 2005 Can desert dust explain the outgoing longwave radiation anomaly over the Sahara during July 2003? *Journal of Geophysical Research,* **110**, doi:10.1029/2004JD005232.

[27] Prospero, J. M., 1999 Long-term measurements of the transport of African mineral dust to the southeastern United States: Implications for regional air quality. *Journal of Geophysical Research,* **104**, D13, 15917-15927.

[28] Moulin et al., 1997 Control of atmospheric export of dust from North Africa by the North Atlantic oscillation. *Nature*, **387** (6634), 691-694.

[29] Ginoux, P., J. M. Prospero, O. Torres and M. Chin, 2004 Long-term simulation of global dust distribution with the GOCART model: correlation with North Atlantic Oscillation. *Environmental Modelling & Software*, **19**, 113-128.

[30] Pérez C., P. Jiménez, O. Jorba, J. M. Baldasano, E. Cuevas, S. Nickovic, X. Querol 2007 Long-term simulations (1958-2006) of Saharan dust over the Mediterranean and the Eastern North Atlantic with the DREAM regional dust model. IUGG 2007 Perugia, Session: MS004 Mineral Dust Cycle and its Impact on Clouds and Radiation (ICCP), Perugia (Italia), 2-13, July.

[31] Pérez, C., Haustein, K., Janjic, Z., Jorba, O., Baldasano, J.M., Black, T., Nickovic, S. 2008 An online mineral dust model within the global/regional NMMB: current progress and plans. AGU Fall Meeting. San Francisco, 15-19 , December 2008.

IOP Conf. Series: Earth and Environmental Science **7** (2009) 012014 doi:10.1088/1755-1307/7/1/012014

LMDzT-INCA Dust Forecast Model Developments and Associated Validation Efforts

M Schulz, A Cozic and S Szopa

Laboratoire des Sciences du Climat et de l'Environnement, CEA-CNRS-IPSL, Gif-sur-Yvette Cedex, France

Email: michael.schulz@lsce.ipsl.fr

Abstract. The nudged atmosphere global climate model LMDzT-INCA is used to forecast global dust fields. Evaluation is undertaken in retrospective for the forecast results of the year 2006. For this purpose AERONET/Photons sites in Northern Africa and on the Arabian Peninsula are chosen where aerosol optical depth is dominated by dust. Despite its coarse resolution, the model captures 48% of the day to day dust variability near Dakar on the initial day of the forecast. On weekly and monthly scale the model captures respectively 62% and 68% of the variability. Correlation coefficients between daily AOD values observed and modelled at Dakar decrease from 0.69 for the initial forecast day to 0.59 and 0.41 respectively for two days ahead and five days ahead. If one requests that the model should be able to issue a warning for an exceedance of aerosol optical depth of 0.5 and issue no warning in the other cases, then the model was wrong in 29% of the cases for day 0, 32% for day 2 and 35% for day 5. A reanalysis run with archived ECMWF winds is only slightly better (r=0.71) but was in error in 25% of the cases. Both the improved simulation of the monthly versus daily variability and the deterioration of the forecast with time can be explained by model failure to simulate the exact timing of a dust event.

1. LMDZT-INCA aerosol model

The global aerosol transport model LMDzT-INCA is operated for an experimental chemical weather forecast system. The model includes a dust aerosol component as part of a multi-component aerosol module. For the forecast version we nudge the winds, which means that three dimensional wind fields for every 6 hour time step for all the forecasted time period are used from NCEP and ECMWF to guide the model wind fields and hence force the atmospheric model LMDzT to follow synoptic weather evolution. The relaxation of the GCM wind components u and v toward NCEP or ECMWF meteorology is performed by applying a correction term to the GCM winds with a relaxation time of 2.5 h for both variables and over the whole model domain. For experimental purposes we use three day forecasted winds for NCEP and five days from ECMWF. We use the standard LMDzT-INCA model as described e.g. by [1] and its aerosol module as documented within the AeroCom model intercomparison effort [2] with a horizontal resolution of ca 3.75° x 2.85°.

Dust erosion by wind is responsible for the largest fraction of the global aerosol mass load. Erosion takes place mainly in semi-arid and arid regions but not everywhere in such regions and not at all times. While the erosion of dust is a strong function of wind speed, approximately to the power of three, it is necessary to account in a dust source model for limitations in erodibility. We have

© 2009 IOP Publishing Ltd

implemented several such limitations: the "*erosion potential*" of a given region provides the maximum amount of dust which is effectively available for atmospheric transport under favourable wind speed conditions; the "*varying state of the soil*" can introduce a variability in the erosion potential through wetting of the soil and freezing conditions; the "*threshold wind speed*" expresses that limit above which wind stress can effectively start the erosion process; the initial "*dust size distribution*" corresponds to that fraction of the saltated soil particle spectrum which is long range transported. The simplified formula of the INCA dust flux F_{dust} reads as follows (equation 1):

$$F_{dust} = C \cdot u^2 \cdot (u - u_t) \cdot F_{soil} \qquad \text{if} \qquad u > u_t \qquad (1)$$

where C represents the erosion potential, u the surface wind speed, u_t the threshold velocity above which wind erosion can take place, F_{soil} represents the inhibiting factor due to unfavorable soil conditions. The erosion potential C is varied for different dust minerals and for number and mass of uplifted dust to provide an initial size distribution. We assume, that for a given sufficiently high wind speed, in dry conditions, a constant maximum erosion potential is valid for each region. This potential will integrate subgrid scale variability of erodibility and vegetation cover. It represents the effective grid area fraction in which the soil is erodible. It also incorporates a parameterisation for which fraction of the saltated dust leaves the close-to-the-ground saltation layer and enters into the atmospheric boundary layer for transport. We have chosen to use the remote sensing products from TOMS and Aeronet to obtain an erosion potential C for larger regions [3,4]. It is defined only in regions where precipitation is less than 300 mm per year. Values close to 2 ug m^{-5} s^{-2} are retrieved, but vary between 0.05 and 2 for the different deserts of the world. On rare occasions it rains also in arid and semi-arid regions and the state of the soil becomes unfavourable to dust erosion. We assume that a wet or frozen soil inhibits dust erosion, since soil particles are bound together with water or ice. [5] and [6] have shown that corn size and temperature are the most important factors to explain how long water is retained in top desert soils. Dust is known to be eroded only above a threshold velocity which may vary from region to region. The INCA dust source extends the mapped Saharan dust source of [7] [MB95]. This is done through the mineralogical information for arid and semi-arid areas contained in the FAO (Food Agriculture Organisation) database of soil types [8,9], and by establishing a spatial correspondence between threshold velocities from MB95 and the different FAO soil types. The residence time of mineral dust in the atmosphere depends to a large extent on the initial size distribution of the dust, since sedimentation is the major removal process near the sources. Our sensitivity tests in the chemical transport model TM2 [10] showed that an initial size distribution with a mass median diameter of 2.5 μm with a geometric standard deviation of two is a good assumption to be able to simulate the long range dust transport of dust. At the same time note that the modal approach used in the LMDzT-INCA aerosol module assures that the size distribution evolves with time of transport.

2. Definition of Experimental runs

The experimental forecast system is running since summer 2005. It has been originally built to support experimental campaigns. Daily forecasted NCEP and ECMWF winds are retrieved once a day to nudge the LMDzT model. Each daily forecast simulation encompasses a period of three days (based on NCEP winds) or five days (based on ECMWF winds). While the nudging is done at the model resolution of 3.75°x2.85°, forecasted surface winds are used in addition at a higher resolution of 1x1 degree to compute with higher accuracy the dust fluxes. Dust fluxes are then interpolated to the transport model resolution at every time step (30 minutes). Results have been stored from all model runs and we examine exemplary our simulations forced by ECMWF winds. Results are here combined into three continuous time series of simulated aerosol fields, corresponding to the initial day of each simulation (named J0) and to the forecasted day 2 and 5 of each simulation. Model runs are performed every day based on the restart file obtained at the end of the first day of yesterday's simulation. The restart files contain the initial conditions of the meteorological parameters and tracer mixing ratio

IOP Conf. Series: Earth and Environmental Science **7** (2009) 012014 doi:10.1088/1755-1307/7/1/012014

fields. Note that only the winds are taken as external forcing from NCEP and ECMWF for a suite of forecast simulations. Once the forecast chain is broken it requires several days before the system is at equilibrium with respect to the aerosol fields. In the case of a discontinuation of the model runs, the latest available restart file is used. The results of the forecast model with respect to the evolution of the aerosol and gas species are accessible via a web interface: http://www-lsceinca.cea.fr/cgi-bin/lsce/inca_work_annualrs.pl. A reanalysis run was performed later which made use of the operational ECMWF winds. This run did not experience interruptions as did the experimental forecast system.

3. Evaluation of the LMDZT INCA dust forecasts

Validation is done here on an exemplary basis using Aeronet sun photometer and MODIS satellite derived aerosol optical thickness at 550 nm (AOD). Figure 1 shows the comparison of a dust cloud exiting North Africa on September 4, 2005, observed by MODIS and compared to the simulated AOD field of the INCA forecast model version. The form of the dust plume in both satellite observation and modeled dust AOD field suggests that the dust had exited across a small stretch of the African coast. In both images one can also see the dispersion of the dust over the Atlantic, which involved a southward directed vortex. Such correspondence between model and satellite has been found on several occasions.

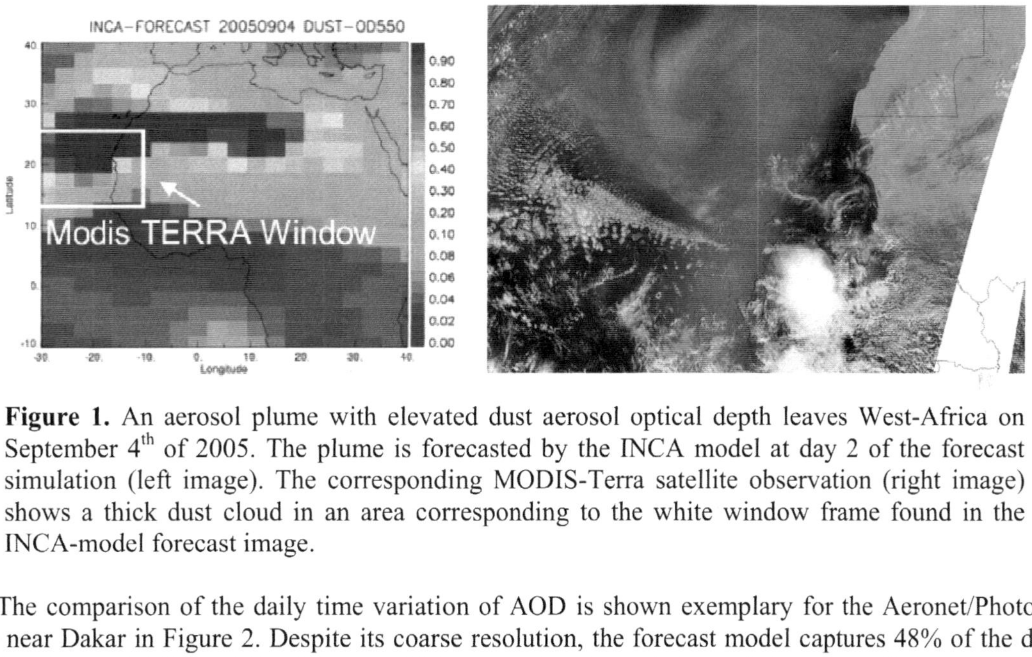

Figure 1. An aerosol plume with elevated dust aerosol optical depth leaves West-Africa on September 4th of 2005. The plume is forecasted by the INCA model at day 2 of the forecast simulation (left image). The corresponding MODIS-Terra satellite observation (right image) shows a thick dust cloud in an area corresponding to the white window frame found in the INCA-model forecast image.

The comparison of the daily time variation of AOD is shown exemplary for the Aeronet/Photons site near Dakar in Figure 2. Despite its coarse resolution, the forecast model captures 48% of the day to day dust variability within the year 2006 near Dakar (reanalysis model version 50%). On weekly and monthly scale the model captures respectively 63% and 71% of the variability (reanalysis version: 70% and 74%). Most of the smaller dust events are found both in the Aeronet and modeled time series. For some large dust events the coherence is less well. The results suggest that for some large dust events the model seems to predict too high dust levels. In the reanalysis run, this feature of the model disappeared partly (not shown). Since the dust emissions and thus the amplitude of the dust events is largely driven by the surface winds, which are directly taken from NCEP and ECMWF for our forecast model, we suspect that there is a major difference in the reanalysis ECMWF surface wind fields.

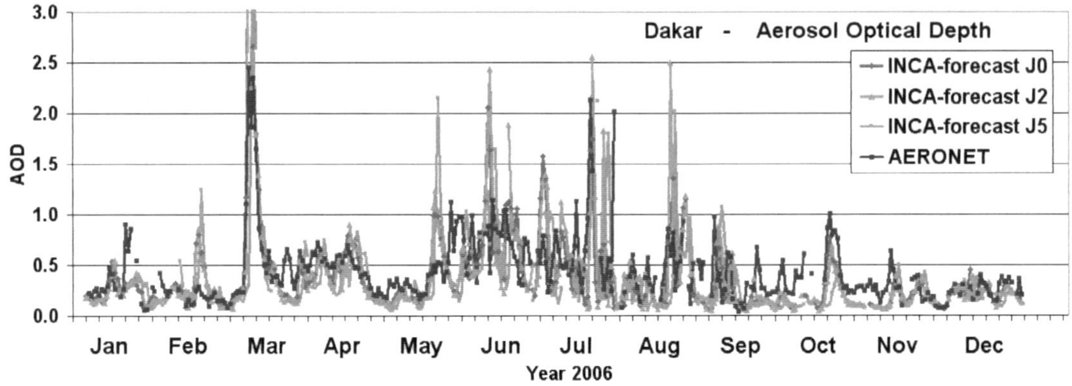

Figure 2. Mean daily total aerosol optical depth as measured at the AERONET/PHOTONS site near Dakar in West Africa by a sun photometer against the INCA model simulated values. Three modelled time series are shown: J0 corresponds to the day when the model is initialised, J2 corresponds to day two of the forecast and J5 uses the forecast five days ahead.

We explore also the deterioration of dust forecast with time of forecast. Correlation coefficients between daily AOD values observed and modelled at Dakar decrease from 0.69 for the initial forecast day J0 to 0.59 and 0.41 respectively for two days ahead and five days ahead. The correlation found for the reanalysis run is found to attain 0.71.

Other possibilities to judge the forecast model are hit rates and the error in predicting high levels of dust. Hit rates are established by calculating those occurrences when the model AOD level is within a range of 0.3 of the observed value. Such hit rates attain 76%, 72% and 64% for the time steps J0, J2 and J5. This shows that the model can predict major dust events, which can result in high AOD values of well above 1 in the area as can be seen from figure 2. We also calculate the possible error in eventually issuing a warning for major dust events. We define here a major dust event as one with AOD values greater 0.5. Table 1 reports the cases when the model would have issued a correct or a false dust warning, in other words when it would have missed an event and when it correctly assumed that dust is below the warning level. For a threshold level of 0.5 AOD we find that at Dakar the model is correctly predicting on 71% (68%, 65%) of the days for time step J0 (J2, J5 respectively) the state of the dust load, be it over the threshold or below. As can be seen from the detailed error analysis in Table 1, an AOD greater than 0.5 is observed on 36% of the days. These dusty days are predicted correctly almost every third time even on forecast day five, when 27 events are correctly predicted and 52 are missed out. However there are also quite some false alarms (or falsely timed alarms). Even for forecast day 5 this error frequency is only 35%.

Table 1. Number of days in 2006, when the INCA model would be right/wrong with respect to predicting a dusty situation with levels above 0.5 AOD near Dakar. The error counts are based on the comparison of the daily Aeronet and INCA modelled AOD levels. Three time steps of the forecast model run and the reanalysis run are shown. Days with no information from Aeronet or from the forecast model are omitted.

#	J0	J2	J5	ReAnalysis
missing « warning »	43	47	52	40
correct « no warning »	118	116	114	124
correct « warning »	36	32	27	39
« warning » false	21	23	25	15

A second time series comparison is shown in Figure 3. From the model we can retrieve the contribution of dust to total AOD. It can be seen that for both the dusty cases and the mixed dust/pollution cases model data correlation in daily AOD evolution can be found.

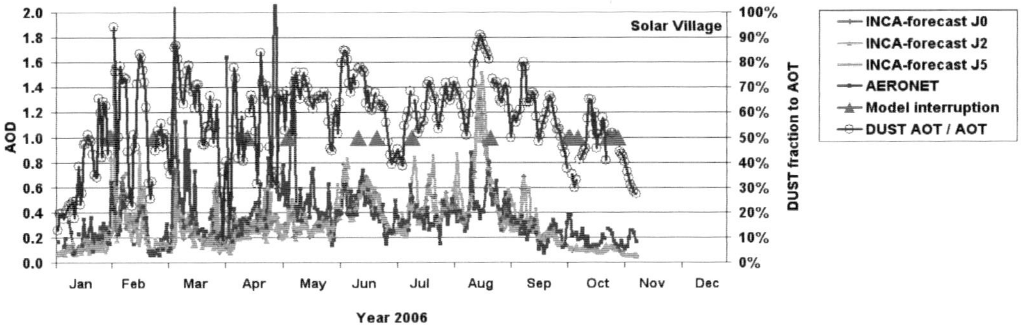

Figure 3. Aeronet sun photometer measurements of aerosol optical depth (AOD) at Solar Village in Saudi Arabia as compared to day zero (J0) day two and five (J2,J5) of all daily INCA forecast simulations performed in the year 2006. Pink triangles indicate times when model operation was discontinued. The contribution of dust AOD to total AOD from the model is shown as brownish curve.

4. Conclusions

The forecast simulation is a relative harsh test of modeling capabilities, since the wind fields, the initialization and the dust aerosol source formulation undergo no further checking. Despite this the INCA model shows significant capacity in predicting dust events in West Africa.

The comparison of the monthly versus the daily model/data correlation and the deterioration of the forecast with time can be explained by the models failure to simulate some aspects of the transport and thus the exact timing of a dust event.

The model agreement with observed AOD at dusty sites as a function of forecast time and on different temporal scales, hit rates and an error analysis with respect to threshold dust loads is suggested to be one criterion for evaluating the quality of the transport of dust in a given model.

Acknowledgments

Of great benefit have been the Aeronet/Photons data of sun photometers provided through the Aeronet web site, namely Brent Holben and Didier Tanre. The operational forecast model version is based on tools developed at the LMD, thanks to Frederic Hourdin. Support by the supercomputing facilities of the CEA and the IT department at the LSCE, Francois Marabelle, is acknowledged. Financial support has been received through the EU funded GEMS project.

References

[1] Hauglustaine, D.A., F. Hourdin, L. Jourdain, M.-A. Filiberti, S. Walters, J.-F. Lamarque, and E.A. Holland, 2004 Interactive chemistry in the Laboratoire de Météorologie Dynamique general circulation model : Description and background tropospheric chemistry evaluation, *Journal of Geophysical Research*, **109** (D04314, doi:10.1029/2003JD003957)

[2] Textor, C., M. Schulz, S. Guibert, S. Kinne, Y. Balkanski, S. Bauer, T. Berntsen, T. Berglen, O. Boucher, M. Chin, F. Dentener, T. Diehl, R. Easter, H. Feichter, D. Fillmore, S. Ghan, P. Ginoux, S. Gong, A. Grini, J. Hendricks, L. Horowitz, P. Huang, I. Isaksen, T. Iversen, S. Kloster, D. Koch, A. Kirkevåg, J.E. Kristjansson, M. Krol, A. Lauer, J.F. Lamarque, X. Liu, V. Montanaro, G. Myhre, J. Penner, G. Pitari, S. Reddy, Ø. Seland, P. Stier, T. Takemura, and X. Tie, 2006 Analysis and quantification of the diversities of aerosol life cycles within AeroCom, *Atmos. Chem. Phys.*, **6**, 1777-1813.

[3] Balkanski, Y., M. Schulz, T. Claquin, C. Moulin, and P. Ginoux, 2004 Global emissions of mineral aerosol: formulation and validation using satellite imagery, in *Emission of Atmospheric Trace Compounds*, edited by C. Granier, P. Artaxo, and C.E. Reeves, pp. 239-267, Kluwer, 2004.

[4] Claquin, T., 1999 Modelisation de la minéralogie et du forcage radiatif des poussières désertiques, PhD thesis, Université Paris VI, France, Paris.

[5] Shao, Y., M. Raupach, and J. Leys, 1996 A model for predicting aeolian sand drift and dust entrainment on scales from paddock to region, *Australian Journal of Soil Research*, ***34*** (3), 309 - 342.

[6] Fecan, F., B. Marticorena, and G. Bergametti, Parametrization of the increase of the aeolian erosion threshold wind friction velocity due to soil moisture for arid and semi-arid areas, *Annales Geophysicae-Atmospheres Hydrospheres and Space Sciences*, **17** (1), 149-157.

[7] Marticorena, B., and G. Bergametti, 1995 Modeling the atmospheric dust cycle: 1. Design of a soil-derived dust emission scheme, *J. Geophys. Res.*, **100**, 16415-16430.

[8] FAO, 1995 Digital soil map of the world, Food and agriculture organization of the United Nations (FAO).

[9] Zobler, L., 1986 A world soil file for global climate modeling, pp. 32, NASA, Washington, D.C.

[10] Schulz, M., Y.J. Balkanski, W. Guelle, and F. Dulac, 1998 Role of aerosol size distribution and source location in a three-dimensional simulation of a Saharan dust episode tested against satellite-derived optical thickness, *Journal of Geophysical Research-Atmospheres*, **103** (D9), 10579-10592.

IOP Conf. Series: Earth and Environmental Science **7** (2009) 012015 doi:10.1088/1755-1307/7/1/012015

African Dust Deposition to Florida: How well do dust models perform?

J M Prospero[1] and W Landing[2]

[1] Rosenstiel School of Marine and Atmospheric Science, University of Miami. Miami, USA.
[2] Department of Oceanography, Florida State University. Tallahassee, USA.

E-mail: jprospero@rsmas.miami.edu

Abstract. Models, widely used to estimate dust deposition to the oceans, have not been adequately tested because of the dearth of long-term measurements of dust deposition rates. Here we present the results of a three year deposition program in a nine-station network distributed throughout the State of Florida which is impacted by African dust every year. The deposition rates of soil-related elements (Al, Fe, Mn) were strongly correlated and exhibited an extremely strong summer maximum that closely matched concurrently-measured dust concentrations in Miami. The deposition rates of Al, Fe and Mn at all sites were remarkably uniform. The deposition rates of dust, based on Fe measurements, were compared with those from nine AEROCOM models. Dust models correctly characterized the seasonal cycle of deposition but most grossly underestimated summer deposition and they tended to produce latitudinal gradients that were not observed in the network. There is a clear need to better understand the treatment of deposition processes in models.

1. Introduction

Models are widely used to assess dust-Fe transports to the global ocean [1]. While they agree in a general way on gross transports, they often yield strongly divergent results for deposition to specific basins [2]. While considerable efforts have focused on the source function and transport process, removal processes – by various wet and dry deposition mechanisms – have received little attention. A major problem is that there have been very few measurements of dust deposition in marine environments.

Here we present the results of an extensive series of wet deposition measurements made in Florida over a three year period [3] in a network of nine stations that extended from the Florida Keys through central Florida, to the Florida Panhandle, spanning approximately six degrees of latitude and five of longitude. Florida is heavily impacted by African dust every year [4,5]. African dust is readily measured in wet deposition; in some cases dust is clearly visible as a red-brown "mud" in the collector [6]. Here we report on the temporal-spatial variability of dust deposition in the network and we interpret these data in terms of the long-term measurements of dust made at a University of Miami coastal site on an island east of Miami [4]. Finally we compare these dust measurements to various AEROCOM model products that estimate dust deposition rates. We address three major questions:

© 2009 IOP Publishing Ltd

- How much African dust is deposited in Florida each year?
- How does the quantity of deposited dust vary in time and space throughout the State?
- How well do dust transport-deposition models perform in replicating the temporal-spatial pattern and amounts of deposition?

Month-long rainfall samples were collected in the Florida Atmospheric Mercury Study (FAMS) network (figure 1) [7]. To minimize impacts from local sources, sampling was carried out atop a 15m tall aluminum scaffold tower using automatic wet-dry collectors and bulk (permanently open) collectors [8,9]. In this report the term "wet" deposition (WD) refers to the samples collected in the "wet" bucket in the automatic sampler and "bulk" deposition (BD) refers to the samples collected in the permanently-open funnels.

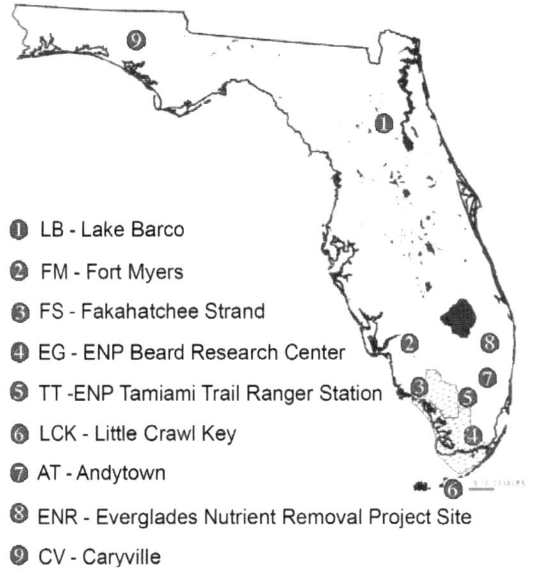

Figure 1. Map of FAMS Network Stations

2. Results

2.1. Elemental Concentrations in Wet Deposition: Natural and Anthropogenic Sources

We would expect the elements Al, Fe, and Mn to be most unambiguously representative of the dust component in precipitation because of their high concentration in crustal materials (8.04%, 3.5% and 0.06% respectively; [10]). Scatter plots of Fe vs. Al in wet deposition showed strong correlations across the entire network. Although the concentrations of Fe and Al are highly correlated at all sites, the slopes of the regression lines are about twice that of the crustal ratio, 0.43. Similarly, scatter plots of Mn vs. Al yielded very solid regression lines but the slopes, average 0.022, were about three times the ratio of Mn to Al in crustal material, 0.0075 [10]. In contrast, scatter plots of Mn vs. Fe yielded strong regressions; slopes averaged 0.022, close to that of average crust, 0.017. The high ratios of Fe/Al and Mn/Al relative to crust and the close ratio of Mn/Fe are attributed to the difficulty of measuring Al with ICPMS procedures with the instrumentation available at that time. In contrast to Fe, Al, and Mn, pollution elements show a very different behavior; they are uncorrelated with Fe and their ratios to Fe are much higher than that in average crustal material.

2.2. Iron Wet Deposition and African Dust Aerosol Concentrations

Figure 2, top, shows the monthly Fe WD and rainfall collected at Tamiami Trail along with atmospheric dust concentrations measured daily at a coastal site on Virginia Key, 66 km E of the TT site (for details, see [4]). The notable feature in this figure is that the seasonal pattern of Fe WD closely follows that of the dust measurements at the coastal site. It is also notable that while the rainy season extends well beyond the African dust season there is little or no Fe WD in September and October, months that are typically quite rainy but when dust concentrations are low. The picture is different for pollution elements Pb, Cu, and V (figure 2, bottom) which are deposited over more months than Fe because of the year-long presence of pollutants which are assumed to be mainly derived from local and regional sources.

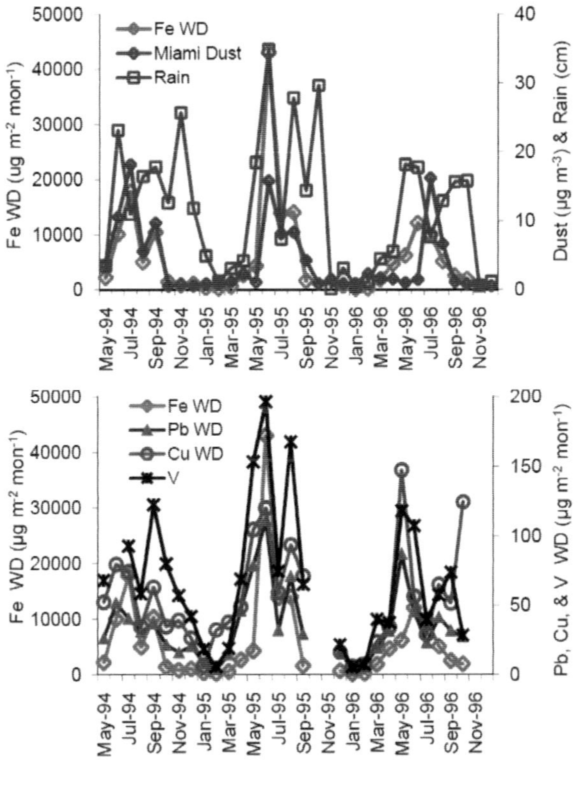

Figure 2. Tamiami Trail Wet Deposition and African dust measured in Miami. Top: Monthly values of TT Fe WD, TT rain, and Miami dust. Bottom: Monthly values of TT WD of Fe, Pb, Cu and V.

2.3. Dust Deposition Rates

Henceforth, all subsequent discussion will focus on dust deposition rates calculated from the Fe concentrations ($Dust_{Fe}$) assuming an average crustal abundance of 3.5%. The monthly $Dust_{Fe}$ deposition rates in the entire network of nine stations are shown in figure 3. A number of features stand out. First, there is clearly a very strong seasonality that is broadly consistent across the three years of record; in all three years the maximum deposition months fall in June, July or August. Second, with some notable exceptions, during the dust season $Dust_{Fe}$ WD rates fall in a relatively narrow range, roughly 20 to 50 $\mu g\ cm^{-2}\ mon^{-1}$. Third, sporadically at some stations unusually high WD rates are obtained: e.g., in July 1994 at LB, in June 1996 at TT, and in the summer of 1996, at four stations in South Florida. These spikes in deposition demonstrate the great variability of scavenging events, requiring as they do the chance occurrence of high concentrations of dust with a strong rain event.

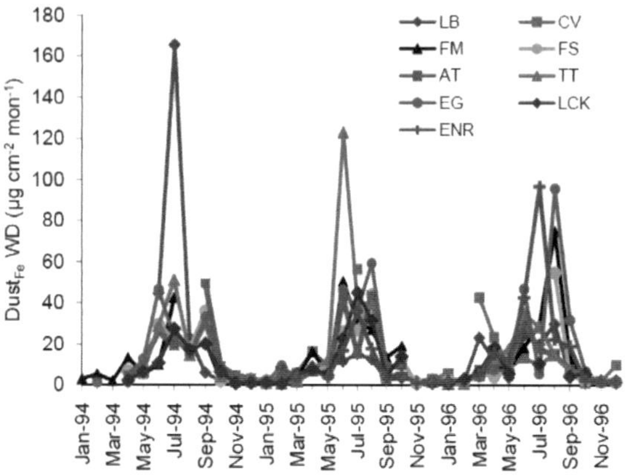

Figure 3. Monthly Dust$_{Fe}$ WD in the FAMS network, 1994 – 1996.

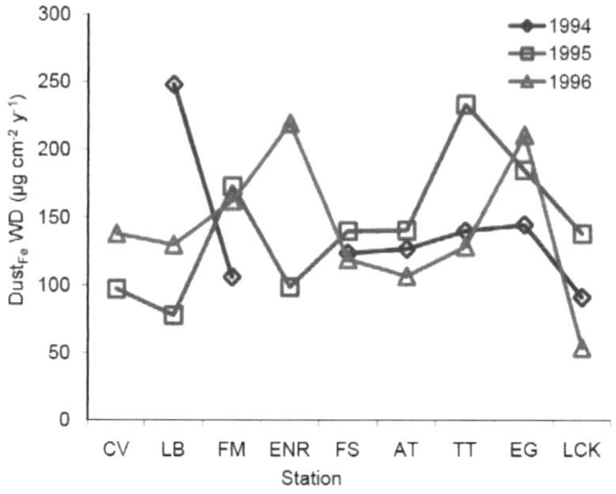

Figure 4. Annual mean values of Dust$_{Fe}$ WD at each of the FAMS network stations in each year.

Figure 3 suggests that dust deposition rates were relatively uniform across the entire FAMS network from year to year. This is supported by figure 4 which shows the annual Dust$_{Fe}$ WD rates at each station in the network in each of the three years. It is clear that WD rates are relatively uniform across the network. In any one year, WD rates mostly fall in a band between 100 and 250 µg cm^{-2} mon^{-1}. Also there is no evidence of a systematic trend in the concentrations that would suggest, for example, that more dust is deposited in South Florida than in North. The uniformity of Dust$_{Fe}$ WD over the entire latitudinal span of the network is clearly seen in figure 5 which shows the mean annual WD and BD at each site over the entire program. Excluding LCK, where climatological rainfall is substantially lower than the other eight sites, the annual average deposition fell within a relatively narrow range, roughly 120 to 180 µg cm^{-2} yr^{-1} and BD between about 180 to 260 µg cm^{-2} yr^{-1}.

2.4. Wet Deposition Compared to Bulk Deposition

WD rates were highly correlated with BD rates. At all sites scatter plots of Fe, Al, and Mn monthly WD vs. BD yielded excellent regression lines with slopes between 0.74 to 0.80 and R^2 values over 0.9. (The apparent ratios in figure 5 are more variable than the regression ratios because figure 5 is based on all data which includes unpaired WD and BD values.) At low deposition rates there is a clear tendency for data to fall to the right of the regression line, evidence that dry deposition is dominant. This is not surprising given the generally small amounts of rain during the dry season (i.e., the winter and spring) when increased amounts of local dust might be expected. Nonetheless it is clear that, in general, local sources do not contribute substantially to the total deposition of soil dust to the bulk deposition collector.

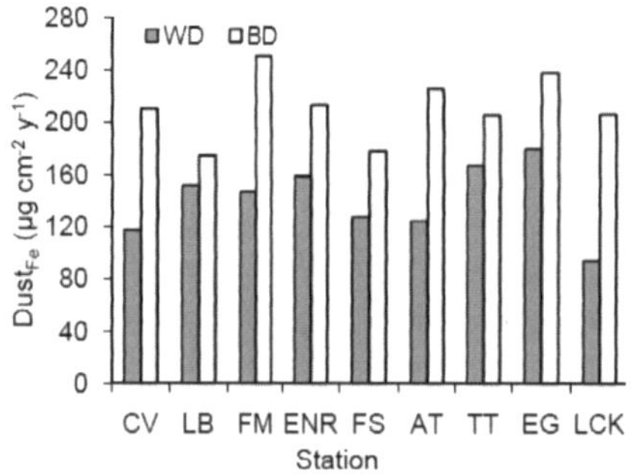

Figure 5. Annual average Dust$_{Fe}$ WD and BD at each network station over 1994 – 1996.

2.5. Comparison of Model Deposition with FAMS Measurements

Here our objective is to compare the performance of various models with our deposition measurements. We wished to compare the differences between models and to look for systematic trends in the models as compared to our measurements. Unfortunately there was little modeling data available for the years 1994 – 1996. As a preliminary effort we used AEROCOM Chemical Transport Model data (http://nansen.ipsl.jussieu.fr/AEROCOM/data.html) for the year 2000, the only year with a large model set. Data from nine models were extracted (by Michael Schulz, CEA/CNRS) for five 1° by 1° boxes that encompassed one or more of the nine FAMS stations. We caution that the results are not directly comparable because of the different time spans. However rainfall amounts and dust concentrations in 2000 were similar to those in 1994 – 1996; consequently we might not expect to see wide divergences due to these factors. Nonetheless, given the difference in years, we do not explicitly identify models; we simply identify them by numbers.

In figure 6 we show the $Dust_{Fe}$ WD measured at TT with the model 2000 data. We replicate the model data over the three years of the FAMS data so that we can observe the variability of the measurements relative to the repeated model pattern. Most models did a good job of capturing the seasonal cycle in WD but six of the models grossly underestimated the summer deposition amounts, producing only a modest deposition peak. This pattern was repeated at all sites with usually the same models yielding relatively low WD rates. However many models showed a gradient over the length of the State with lower WD rates in the northern regions, a trend not seen in the FAMS data.

The low dust deposition produced by many of the models is clearly seen in figure 7 which compares the WD and BD yielded by the models with that measured in the four stations located in the one-degree grid box in the southern end of the State: TT, FS, AT, EG. The differences would have been greater if we had only extracted the summer months for the comparison. Similar comparisons for the other sites in Central and North Florida and at LCK yielded similar results.

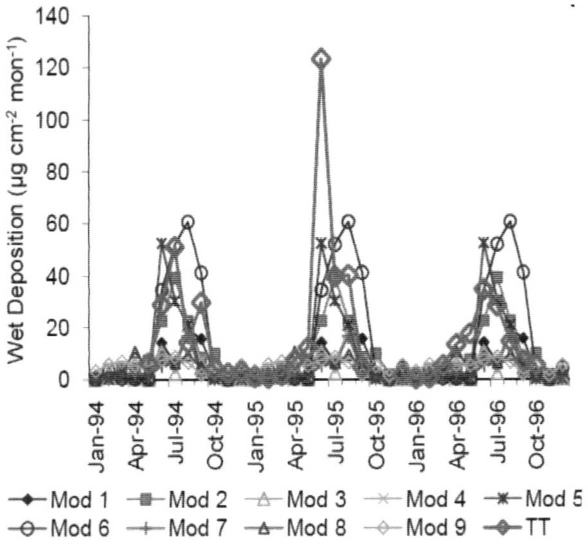

Figure 6. Monthly $Dust_{Fe}$ WD at TT along with estimated WD in nine AEROCOM models.

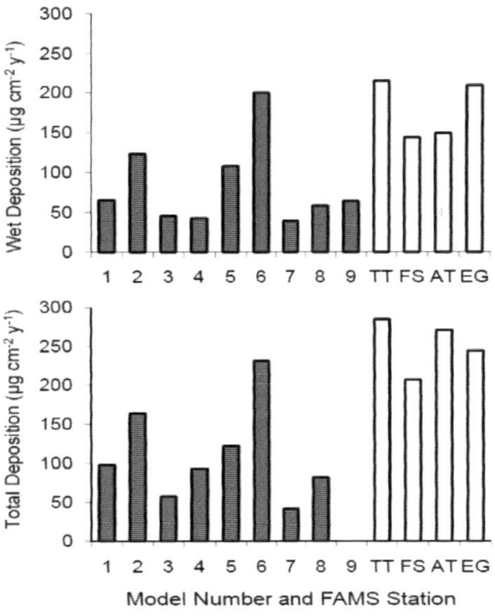

Figure 7. Comparison of modelled and measured Dust Wet Deposition and Total Deposition for the one- degree square containing four FAMS stations: TT, FS, AT, EG.

3. Conclusions

The $Dust_{Fe}$ WD and BD rates were remarkably uniform over the length of Florida, about 150 and 200 μg cm^{-2} yr^{-1}, respectively. The WD rate is quite similar to that obtained in a one year study (1982 – 1983) of daily dust deposition at the UM coastal site in Miami, 126 μg cm^{-2} yr^{-1}. These rates, roughly 1 to 2 g m^{-2} yr^{-1}, are geologically significant. They are comparable to the non-authigenic deep-sea sediment accumulation rates in the tropical Atlantic. Also African dust is a significant contributor to soils in the western Atlantic. A recent study [11] shows that African dust is the dominant source of the clay fraction of soils on Barbados, the Florida Keys and the Bahamas Islands.

Our study shows that we need a better understanding of the wet and dry removal processes that act on dust particles. To the extent that our model comparison is valid, given the gap in dates, the wide disparity among the model results suggests that there are major differences in the models. However it is not possible to ascribe these differences to specific modeling aspects of sources, transport, or removal. While the source and transport performance of models is often tested with remote sensing data (e.g., aerosol optical depth measured by satellite and ground measurements such as those in AERONET) there is no quantitative data set that can be used to validate deposition estimates to ocean regions. We need high quality deposition data for this purpose. This task will require considerable effort because of the variability of dust deposition events in Miami [6], 22% of the total annual dust deposit occurred in one day and 68% in two dust events spread over a four-day period. Thus extended time-series studies are needed. On the other hand, the uniformity of dust deposition across the length and breadth of the State suggests that relatively few stations are required to document deposition. Clearly, we need much more deposition data from other locations impacted by dust. While island locations are preferable, measurements in continental-marine areas can be useful as shown in this study if precautions are taken to characterize the possible impact of local sources.

Acknowledgements

We thank Michael Schulz, Laboratoire des Sciences du Climat et de l'Environnement CEA/CNRS – LSCE, for assistance in extracting AEROCOM model data for our study, and Jerome J. Perry for the ICPMS analyses. This work was supported in part by National Science Foundation grant OCE-0623189 and by funding for FAMS from FDEP and EPRI.

References

[1] Jickells, T. D., et al. 2005 Global Iron Connections Between Desert Dust, Ocean Biogeochemistry, and Climate, *Science*, **308**(5718), 67-71.

[2] Mahowald, N. M., et al. 2005 Atmospheric global dust cycle and iron inputs to the ocean, *Global Biogeochem. Cycl.*, **19,** GB4025, doi:10.1029/2004GB002402.

[3] Landing, W. M., et al. 1995 Relationships between the atmospheric deposition of trace elements, major ions, and mercury in florida: The FAMS project (1992–1993), *Water Air Soil Pollu.*, **80**(1), 343-352.

[4] Prospero, J. M. 1999 Long-term measurements of the transport of African mineral dust to the southeastern United States: Implications for regional air quality, *J. Geophys. Res.*, **104**(D13), 15917-15928.

[5] Prospero, J. M., et al. 2001 Al and Fe in PM 2.5 and PM 10 Suspended Particles in South-Central Florida: The Impact of the Long Range Transport of African Mineral Dust, *Water Air Soil Pollu.*, **125**(1), 291-317.

[6] Prospero, J. M., et al. 1987 Deposition rate of particulate and dissolved aluminum derived from Saharan dust in precipitation at Miami, Florida, *J. Geophys. Res.*, **92**(D12), 14,723-714,731.

[7] Pollman, C., et al. 1995 Overview of the Florida Atmospheric Mercury Study (FAMS), *Water Air Soil Pollu.*, **80**(1), 285-290.

[8] Landing, W. M., et al. 1998 Methods for measuring mercury in rainfall and aerosols in Florida, *Atmos. Environ.*, ***32***(5), 909-918.

[9] Pollman, C. D., et al.. 2002 Wet deposition of phosphorus in Florida, *Atmos. Environ.*, **36**(14):2309-2318.

[10] Taylor, S. R., and S. M. McLennan 1985 The continental crust: Its composition and evolution: Oxford, *England, Blackwell Scientific Publishing, 312.*

[11] Muhs, D. R., et al. 2007 Geochemical evidence for African dust inputs to soils of western Atlantic islands: Barbados, the Bahamas, and Florida, *J. Geophys. Res.*, **112**, D13203, doi:10.1029/2006JD007577.

IOP Conf. Series: Earth and Environmental Science 7 (2009) 012016 doi:10.1088/1755-1307/7/1/012016

Impacts of Sand and Dust Storms on Agriculture and Potential Agricultural Applications of a SDSWS

R Stefanski and M V K Sivakumar

Agricultural Meteorology Division.World Meteorological Organization. Geneva. Switzerland

Email: RStefanski@wmo.int

Abstract. This paper will give an overview of the various impacts of sand and dust storms on agriculture and then address the potential applications of a Sand and Dust Storm Warning System (SDSWS) for agricultural users. Sand and dust storms have many negative impacts on the agricultural sector including: reducing crop yields by burial of seedlings under sand deposits, the loss of plant tissue and reduced photosynthetic activity as a result of sandblasting, delaying plant development, increasing end-of-season drought risk, causing injury and reduced productivity of livestock, increasing soil erosion and accelerating the process of land degradation and desertification, filling up irrigation canals with sediments, covering transportation routes, affecting water quality of rivers and streams, and affecting air quality. One positive impact is the fertilization of soil minerals to terrestrial ecosystems. There are several potential agricultural applications of a SDSWS. The first is to alert agricultural communities farmers to take preventive action in the near-term such as harvesting maturing crops (vegetables, grain), sheltering livestock, and strengthening infrastructure (houses, roads, grain storage) for the storm. Also, the products of a SDSWS could be used in for monitoring potential locust movement and post-storm crop damage assessments. An archive of SDSWS products (movement, amount of sand and dust) could be used in researching plant and animal pathogen movement and the relationship of sand and dust storms to disease outbreaks and in developing improved soil erosion and land degradation models.

1. Introduction

Sand and dust storms are a natural hazard which can affect large areas of continents and can even produce impacts across continents. As several other papers from this workshop have shown, there are impacts on human health and the economies of countries. This paper will provide an overview of the various impacts of sand and dust storms on agriculture and then discuss the potential applications of a SDSWS for agricultural users. In regards to agriculture, sand and dust storms are related to soil erosion and land degradation. There are several climatic factors that are involved in land degradation such as rainfall, floods, droughts, solar radiation, temperature, evaporation, and wind [1]. In the arid and semi-arid zones of the world, it is estimated that 24% of the cultivated land and 41% of the pasture land are affected by moderate to severe land degradation from wind erosion [2]. Wind erosion is a function of weather events interacting with soil properties and past and present land management through its effects on soil structure and vegetation cover. Drylands are particularly vulnerable to wind erosion since they undergo long periods of dry weather with strong winds that regularly occur, the vegetative land cover does not sufficiently protect the soil, and the soil surface is disturbed due to

© 2009 IOP Publishing Ltd

inappropriate management practices. Soil roughness, erodibility, and wetness along with the quantity and orientation of crop residues are parameters that impact the transport of eroded soils. Over-grazing, over-cultivation and destruction of soil by mining, construction, and off-road traffic stimulates and accelerates soil wind erosion in dryland regions. Sand and dust storms are a large scale manifestation of soil and wind erosion.

2. Impacts of Sand and Dust Storms on Agriculture

There are several different kinds of agricultural impacts due to sand and dust storms [3,4]. One of the most direct impacts is the loss of crop and livestock. There is a direct loss of plant tissue as a result of sandblasting by the sand and soil particles. With this loss of plant leaves, there is reduced photosynthetic activity and therefore reduced energy (sugars) for the plant to utilize for growth, reproduction, and the development of grain, fibre, or fruit. If the timing of the sand and dust storms is early enough in the agricultural season, the plant might be able to regrow the lost leaves and the loss in the final crop yield could be relatively minor. However, even in this instance, any regrowth of leaves will still probably result in yield losses. Additionally, the loss of energy for plant growth would also delay plant development and in regions with short growing seasons, this could increase the end-of-season drought risk by move the moisture-sensitive period (reproduction to grain-filling) past the period of favourable rain and result in lower yields and production. If the sand and dust storms occur later in the season, the plant damage will reduce yield during grain development and if it occurs at maturity but before harvest, there will be a direct harvest loss. If there is a large enough deposit of sand or soil material early in the season, the young plant would be buried and possibly killed to lack of sunlight for photosynthesis. Livestock not sheltered from the storm could be directly harmed and any stress from the physical environment to livestock can reduce their productivity and growth [5].

The loss of soil productivity is another agricultural impact to be considered. The loss of topsoil increases soil erosion and accelerates the process of land degradation and desertification. Topsoil contains the most fertile fraction of the soil and contains many of the vital nutrients such as potassium and phosphorous for plant growth and can be transported long distance. This process is of particularly concern since it potentially affects soil resources and therefore crop productivity on a long-term basis, by removing the layer of soil that is inherently rich in nutrients and organic matter. The cost to productivity is difficult to measure but is likely to be quite substantial. [6] developed an index of wind erosion capacity (C) defined as (equation 1):

$$C = \frac{V^3}{2.9(P - E_p)} \tag{1}$$

where V = wind speed at standard observing levels (~ 10 m), m s^{-1}; P = precipitation (mm); and E_p is potential evapotranspiration (mm). As the above equation demonstrates, when soil movement is sustained, the quantity of soil that can be transported by the wind varies as the cube of the velocity. Models indicate that wind erosion will sharply increase above a threshold wind speed.

Other impacts include filling up irrigation canals with sediments, covering transportation routes, reducing temperatures, covering railroads and roads, causing household dust damage, affecting water quality in rivers and streams, affecting air quality, polluting the atmosphere and destroying mining and communication facilities [3]. They accelerate the process of land degradation and cause serious environment pollution and huge destruction to ecology and the living environment. Atmospheric loading of dust caused by wind erosion also affects human health and environmental air quality.

While this loss of productive topsoil has large negative consequences for the land from which it is taken from, this topsoil can benefit areas where precipitates out. [3] cites several examples of mineral dust playing important role in the supply of nutrients and micronutrients to terrestrial ecosystems. Among the examples is the accumulation of sand-dust from the Sahara into Amazon Valley which brings 1-4 kg of phosphate per ha per year. Also, sand and dust from China are the major component of ice crystals which are alkali and neutralize the emergence of acid rain in Japan.

3. Measures to Combat SDS

Wind and soil erosion have been problems for some time across many different regions of the world. After the Dust Bowl in the 1930s in the United States, the Sahel drought in the 1970s and 1980s, and periodic sand and dust storms in China, many strategies and methods were developed that focus on preventing the soil/sand from being eroded and transported. One the best methods is the use of windbreaks or shelterbelts to reducing the impact of wind speed, which would then reduce the amount of soil material from being carried and transported. A windbreak is defined as structures that reduce wind speed and shelterbelts as rows of trees planted for wind protection [7]. For the best wind reduction and greatest downwind influence, the windbreak should be most porous near the ground and the density of the barrier should increase logarithmically with height in accordance with wind speed. Wind reduction is a function of shelter location as well as the height above the plants. The benefits of windbreaks include decreased soil erosion, increased crop yields, reduced potential and actual evapotranspiration; improved internal plant water relations, reduced livestock stress, control of drifting snow, building maintenance and energy savings [3,7,8]. These generalities are subject to variation depending on soil moisture and the benefits may be most dramatic in dry years or under critical moisture shortages. Examples of widespread use of windbreaks include the U.S. Great Plains after the Dust Bowl years in the 1930s [7]; the Rhone Valley in southeastern France, and the Netherlands [9].

Another method of reducing soil erosion is the use of crop residues [4]. Conservation tillage operations or soil mulching increase the cohesion of soil particles and thereby reduce soil erosion. Surface crop residues help stabilize soils by reducing soil water loss and the erosive force of wind. Vertical residues are much more effective that flat surface cover in controlling soil loss by wind. Special farming implements, such as chisel-type ploughs, which permit the cultivation of vegetated surfaces and maintain a rough, well-textured surface could be used for this purpose [3].

4. Potential Agricultural Applications

There are many potential agricultural applications for a SDSWS and the sand and dust storm models. These can be grouped into three categories: tactical, strategic, and research applications (table 1). The tactical applications focus on actions that farmers can take after a crop has been planted and are of a short-term nature (daily, weekly). Specifically, these are the actions that would be related with a near real-time warning or alert system for farmers and agricultural communities. If this system was in place, farmers could take preventive action such as harvesting grain or vegetable crops that are nearing their maturity and that by harvesting them before the storm, some production could still be obtained. If the crop is immature, the only other action that could be undertaken is physically protecting crops in the field. This is impossible for large fields of crops, but could possibly be done for very small plots of high-value crops (i.e. vegetables and fruits). Another preventive action would be to move livestock to a place (i.e. barn, sheltered valleys) where the impacts of sand and dust storm would be minimized. Finally, with enough warning, farmers or agricultural communities could strengthen infrastructure such as houses, roads, and crop storage facilities before the storm. This would also reduce the physical damage of the storm.

These tactical applications would be the primary focus of a SDSWS. For other applications, the model output of a SDSWS would have greater utility. Strategic applications of sand and dust storm models would be focused on long-term planning and investments in crops and infrastructure that would use improved climatologies based on sand and dust storm models for a particular region or locality. There are already climatologies available for many regions based on long-term weather observations. By integrating the existing long-term observation with the models, more detailed information could be determined such as the average direction and size of sand and dust storm for a region. This would then be used to improve existing or develop new windbreaks and shelterbelts. Another potential application of the sand and dust storm models would be post-storm crop damage assessments. The sand and dust storm models would assist national governments and international institutions in determining whether crop damage in a particular locality was caused by a sand and dust storm and this would aid prevention efforts.

Table 1. Summary of Potential Agricultural Applications of a Sand and Dust Storm Warning System (SDSWS).

Tactical (short term)	Strategic (long-term)	Research
Near-term warnings for agricultural communities to take preventive action:	Improved SDS climatologies for long-term planning for agricultural communities:	Forecasting locust movement
		Improving Soil / Wind Erosion and Land Degradation Models
- harvesting maturing crops	- Planning windbreaks and shelterbelts (direction, size, etc)	
- sheltering livestock,	- Planning for infrastructure and crops	Plant and animal pathogen movement and the relationship of SDS to disease outbreaks
- strengthening infrastructure (houses, roads, crop storage)	- Post-storm crop damage assessments.	Archive of SDSWS products for forensic use

Since sand and dust storm models take into account the movement suspended soil material, they would also be also useful in forecasting locust movement. With the recent locust outbreak in Africa in 2002-2003, WMO and FAO have been coordinating their efforts to assist national locust control personnel in using meteorological data and forecast models. [10] provides monthly updates of the locust situation and the SDS models are potentially well-adapted to be used for this application. With the advent of the sand and dust storm portal, these models can be evaluated for this purpose.

Another important area of research and potential application for sand and dust storm models and SDSWS products is the movement of plant and animal pathogens and the relationship of sand and dust storms to disease outbreaks. Since pathogens can be carried by dust particles and can be considered as dust themselves, sand and dust storm models and SDSWS products could aid in determining the transport of the plant and crop diseases. An archive of SDSWS products (movement, amount of sand and dust) could be used in researching plant and animal pathogen movement. [11] provides an excellent overview and history of plant diseases that have been transported across continents. The limited diversity of plant genetics of globally distributed crops causes a problem of uniform susceptibility to disease and the intercontinental dispersal of pathogens may cause crop diseases on a global scale. For example, the original plantation of Arabic coffee in the Americas can be traced to a single bush taken from Java in 1706. [11] and others give several examples of plant diseases that are suspected of being aerially transported vast distances to impact crops on other continents. These include: sugarcane rust which is suspected of traveling from Cameroon to the Dominican Republic in June 1978 (also [12]); coffee leaf rust from Angola to Bahia, Brazil in 1970; and wheat stem rust from southern Africa to Australia in 1969. Other diseases include the potential recent transport of soybean rust from South America to North America in 2000 and a new form of wheat stem rust that has the potential to move form eastern Africa to South Asia [13]. This transport of pathogens is not limited to plant diseases. There is speculation that Foot and Mouth Disease which affects livestock can be transported by via dust [14,15,16]. Policy makers need to know the source of disease outbreaks whether by airborne or by human transport in order to take preventive measures. An example of this is the uncertainty of the sources of Avian Bird Flu.

Another area of research for the potential agricultural applications of sand and dust storm models and products is in improving soil/wind erosion and land degradation models. For over 100 years, soil erosion data has been collected and analyzed by soil scientists, agronomists, geologists, hydrologists, and engineers and simple soil erosion relationships has been developed from this data that incorporates the major soil erosion factors. The Universal Soil Loss Equation (USLE) was developed in the mid-1960s for understanding soil erosion in agricultural applications and was updated in the 1980s and renamed the Revised Universal Soil Loss Equation (RUSLE) to incorporate the large amount of information that had accumulated since the original. The RUSLE is defined in equation 2 as:

$$A = R \cdot K \cdot L \cdot S \cdot C \cdot P \qquad (2)$$

where A is the soil loss per year (t/ha/year); R represents the rainfall-runoff erosivity factor; K is the soil erodibilty factor; L represents the slope length; S is the slope steepness; C represents the cover management, and P denotes the supporting practices factor [17]. The Wind Erosion Prediction System (WEPS) is a process-based, daily time-step, computer model that predicts soil erosion via simulation of the fundamental processes controlling wind erosion [18]. WEPS can calculate soil movement, estimate plant damage, and predict PM10 emissions when wind speeds exceed the erosion threshold. WEPS is intended for conservation planning, assessing wind erosion and for aiding the development of regional and national policy. Both of these existing models and formulations could be improved by analyzing sand and dust storm model output over many years.

5. Final Thoughts

There are three general components of effective warning systems that a SDSWS should adhere to and these are monitoring, forecasting, and advisories. Potential products for agricultural applications of a SDSWS should include areal extent, severity, duration, and potential impacts. For forensic uses, an archive of SDSWS products (movement, amount of sand and dust) should be developed to be used in researching the relationship of sand and dust storm to plant and animal disease outbreaks and in developing improved soil erosion and land degradation models. Other useful products and information that could be used from sand and dust storm models include insight into weather patterns associated with sand and dust storm, meso-scale features that ignite storms, and the vertical data from the models. Another important aspect is that the various models should be grouped by the capability for users. It is important to note that for operational public warning systems, output from sand and dust storm models are only for guidance and should not be considered as final end-products. This is same issue with Numerical Weather Prediction models used for forecasting day-to-day weather by the National Meteorological Services (NMS) of the world. The end user should not be expected to understand detailed graphs and maps in order to know tomorrow's weather. They rely on weather information that has translated from model output into easy to understand language. Therefore, there is the need for another level between the sand and dust storm models and final users. The NMS of the world could provide this function since they know the local climate and weather patterns and would be best positioned to make interpretations of sand and dust storm models for agricultural decision makers (farmers, agricultural extension agents, government policy makers).

Acknowledgments

We would like to acknowledge the following persons for their assistance and input to this paper: Dr Ray Motha, former President of Commission of Agricultural Meteotology, and Chief Meteorologist at the United States Department of Agriculture (USDA); Gary Peterson, Plant Pathologist at the Agricultural Research Service of USDA; Keith Cressman of the FAO Desert Locust Office in Rome, Italy; and Samsul Huda from the University of Western Sydney, Australia.

References

[1] Sivakumar, M.V.K. and R. Stefanski. 2007 Climate and land degradation: An overview. M.V.K. Sivakumar and N. Ndiang'ui (Eds.). Climate and Land Degradation. Springer-Verlag, Berlin, Germany. 623p.

[2] Rozanov BG 1990 Global assessment of desertification: status and methodologies. In: Desertification revisited: Proceedings of an ad hoc consultative meeting on the assessment of desertification. UNEP-DC/PAC, Nairobi. pp. 45-122

[3] Sivakumar, M.V.K. 2005 Impacts of Sand/Dust Storms on Agriculture. M.V.K. Sivakumar, R.P. Motha, and H.P. Das (Eds.). Natural Disasters and Extreme Events in Agriculture. Springer-Verlag, Berlin, Germany. 367p.

[4] Bielders, C.L., J.L. Rajot, and S. Koala. 1998 Wind Erosion Research in Niger: The experience of ICRISAT and Advanced Research Organizations. Sivakumar, M.V.K., M.A. Zöbish, S. Koala, and T. Maukonen (Eds.). Wind Erosion in Africa and West Asia: Problems and Control Strategies. Proceedings of the ICARDA/ICRISAT/UNEP/WMO Expert Group Meeting 22-25 April 1997, Cairo, Egypt. International Center for Agricultural Research in the Dry Areas (ICARDA), Aleppo, Syria. 198 pp.

[5] Starr, J.R. 1980 Weather, Climate, and Animal Performance. WMO Technical Note No. 190. WMO Publication No. 684. WMO, Geneva, Switzerland

[6] Chepil W.S., Woodruff NP (1963) Physics of wind erosion and its control. *Adv. Agron.* **15**:211-302

[7] Rosenberg, N.J., B.L. Blad, and S.B. Verma. 1983 Microclimate: The Biological Environment. 2nd Edition. John Wiley and Sons, New York, New York.

[8] van Eimern, J., Karschon, R., Razumova, L.A., Robertson, G.W. 1964 Windbreaks and Shelterbelts. WMO No. 147, 188 pp

[9] van Eimern, J., 1968 Problems of shelter planning. Proc. Reading Symp. on Agroclimat. Methods, UNESCO, pp. 157-166.

[10] FAO, 2008. Desert Locust Bulletins. www.fao.org/ag/locusts

[11] Brown J. and M. Hovemoller 2002 Aerial dispersal of pathogens on the global and continental scales and its impact on plant disease. *Science,* **297**. 26 July 2002.

[12] Raloff, J. 2001 Ill Winds: Dust Storms ferry toxic agents between countries and even continents. *Science News.* **160**, No. 14. October 6, 2001.

[13] CIMMYT, 2005 Sounding the Alarm on Global Stem Rust: An assessment of Race Ug99 in Kenya and Ethiopia and the potential for impact in neighboring regions and beyond. By the Expert Panel on the Stem Rust Outbreak in Eastern Africa. http://www.cimmyt.org/english/wps/news/2005/aug /pdf/Expert_Panel_Report.pdf

[14] McKie, R. 2001 Deadly dust 'brought foot and mouth here'. The Observer. September 9, 2001.

[15] Pohl, O. 2003 Disease Dustup: Dust clouds may carry infectious organisms across oceans. Scientific Amercian. June 9, 2003.

[16] USGS, 2003. African Dust Carries Microbes Across the Ocean: Are They Affecting Human and Ecosystem Health? USGS Open-File Report 03-028.http://coastal.er.usgs.gov/ african_dust/ofr-2003-028.pdf

[17] Renard KG, Foster GR, Weesies GA, McCool DK, Yoders CD (1997) Predicting Soil Erosion by Water: A guide to Conservation Planning with the Revised Universal Soil Loss Equation (RUSLE). U.S. Department of Agriculture, Agriculture Handbook No. 703, 404 pp.

[18] Wagner LE (1996) An overview of the wind erosion prediction system. Proc. of International Conference on air Pollution from Agricultural Operations, February 7-9, 1996, Kansas City, MO, pp.73-75. http://www.weru.ksu.edu/weps/docs/wepsoverview.pdf

IOP Conf. Series: Earth and Environmental Science 7 (2009) 012017 doi:10.1088/1755-1307/7/1/012017

Dust and epidemic meningitis in the Sahel: a public health and operational research perspective

M C Thomson[1], I Jeanne[2,3] and M Djingarey[4]

[1] International Research Institute for Climate & Society, Palisades, NY, USA
[2] CERMES/Réseau International des Instituts Pasteur, BP 10 887 Niamey, Niger.
[3] Equipe OTIG, Institut Francilien des Sciences appliqués.
[4] WHO-Multi-Disease Surveillance Centre, Ougadougou, Burkina Faso.

E-mail: mthomson@iri.columbia.edu

Abstract. Now that the health communities attention is increasingly focused on climate-health interactions, it has become essential for health decision makers to better understand the role that climate plays in driving disease burdens and health outcomes (both now and in the future) and the opportunity for integrating climate knowledge and information into health decision-making processes to mitigate the negative and strengthen the positive of climate-health interactions. Here we explore the potential climatic indicators and the climate information needs of relevance to the meningitis prevention and control community engaged in the African 'meningitis belt'.

1. Introduction

"Who ever would study medicine aright should learn the following subjects. First he must consider the effects of the seasons of the year and the differences between them. Secondly, he must study the warm and the cold winds both those are in common to every country and those peculiar to a particular locality. Lastly the effect of water on the health must not be forgotten" (Hippocrates, Father of Medicine, "*Airs, Waters Places*" 460 - 377 B.C).

Despite this ancient knowledge, modern medical technologies emerging in the 1950s obscured for some time the significance of environment and climate as determinants of human disease. The re-emergence of medical geography as a discipline in the 1980's has been driven by the new opportunities created by spatial technologies, including remote sensing and geographical information systems, and has gained momentum through a more recent pre-occupation with our changing climate.

To illustrate the issues we focus on meningococcal (epidemic) meningitis in Sahelian Africa which has emerged in recent years as a disease capable of bringing together a diverse community of health and climate specialists focused on reducing the catastrophic effects caused by this devastating disease. This disease, as its name indicates, is an infection of the fluid that surrounds a person's brain and spinal cord (the meninges). It is one of the most feared diseases in Africa because of its rapid onset and high mortality and morbidity rate; those it doesn't kill often suffer brain damage or deafness. While there are many causes of meningitis, the epidemic form of the disease is caused by bacteria called *Neissera meningitidis*. These bacteria transmitted from person to person by aerosols (e.g. coughing and sneezing), is commonly found in the nose and throat of a significant proportion of the global population where it is usually benign. Under certain circumstances the bacteria becomes

© 2009 IOP Publishing Ltd

pathogenic, invading the naso-pharageal epithelial cells and entering the blood stream – thus instigating disease. Thus changes in the ratio of clinical to subclinical infection, rather than in the risk of infection, explains the seasonal distribution of meningitis, with risk factors for overt disease more extreme in epidemic than in non-epidemic years [1].

Current control measures for the disease involve early detection of an epidemic using an agreed incidence threshold, and the rapid deployment of polysaccharide vaccines once an epidemic alert threshold has been crossed. This strategy is predicated on the fact that the current polysaccharide vaccines lack immunological memory and do not confer herd immunity and therefore have limited value as a routine tool for epidemic prevention but new vaccines are currently being introduced [2].

2. An environmental disease

Meningociccal meningitis is an environmental disease whose spatial and seasonal distribution is readily described by climatic and environmental characteristics [3]. Epidemics of meningococcal meningitis occur throughout sub-Saharan Africa, most frequently in an area, known as the 'Meningitis Belt' [4] that stretches from the Sahelian zone of west Africa to the Horn of Africa. This region is a major source of atmospheric dust over most of north and western Africa and has epidemics and seasonal upsurges in endemic disease in the latter part of the dry season, characterized between November and May by low absolute humidity and the dust-laden Harmattan trade winds. Given the association of the epidemics with a dry and dusty environment and their higher incidence in the 'Meningitis Belt', recent research has focused on developing maps that identify the populations at high risk of epidemics [5] and climate-driven early warning systems that could provide longer lead-times for initiating response [6].

In particular dust has been considered an important candidate for use in an epidemic early warning system as there are numerous reports throughout the literature on the likely involvement of dust in converting the bacteria from its benign form to a pathogenic one [7]. For instance, low absolute humidity and increasing intensity of the Harmattan dust laden winds during an epidemic in northern Nigeria were correlated with a rise in hospital admissions for meningococcal meningitis in the following month. In Benin, a substantial proportion of the temporal variability in cases of disease was attributed to the same factors. The association of epidemics with low absolute humidity and dust is supported by observations that epidemics have been reported following drought years and in dry, dusty conditions, and rarely occur where there is dense forest.

3. The possible role for dust in epidemic meningitis

Although the mechanism by which dust may influence meningitis epidemic occurrence remains unclear the most common explanation for this role is that physical damage to the epithelial cells lining the nose and throat in dry and dusty conditions permits the easy passage of the bacteria into the blood stream [3]. Other potential mechanism include the impact of dust particles on the on fluid dynamics of airborne transmission of the bacteria; the potential impact of dust on preceding viral infection; the widely understood activation of the meningococcus bacteria through the high iron content of dust particles; the impact of high dust levels on human behaviour, including crowding and reduced ventilation (blocking windows etc.) and the impact that dust may have on other climatic variables e.g. absolute humidity and temperature.

Despite the uncertainty surround the potential role of dust in the occurrence of meningococcal meningitis disease one thing is certain: this relationship, if any, is unlikely to be due to the physical carriage of the bacteria on dust particles over long distances as has been suggested elsewhere. All the evidence available to date indicates that the bacteria is transmitted from person to person through aerosols and close contact between individuals is essential for transmission [8].

4. The future

Over the next few years the situation is going to change as the new group A conjugate vaccine, which targets the most important strain of the meningococus bacteria becomes available. This vaccine does confer substantive long term protection and is likely to radically change the epidemiology of meningitis epidemics across the continent and will require a reassessment of the research needs underpinning control activities to facilitate the optimal utilization of resources. However, it is understood that it will take at least 10 years to implement this programme and make accessible the conjugate vaccine to all of those at risk in Africa [9]. While a better understanding of the environmental and climatic determinants of this disease is important for the better targeting of this new vaccine epidemic forecasting remains an important tool for the better use of the current reactive polysacchararide vaccine for the time it will take for the introduction of the new vaccine to be completed.

Forecasting tools require the creation of improved epidemiological and environmental data sets, but equally crucial, is the development of skills for environmental prediction. Key variables for meningitis early warning should occur before or early in the dry season. The vast majority of climate analysis and seasonal climate predictions for Africa has to date focused on the rainy season; now it is time to encourage the climate community to improve their understanding of the African dry season in terms of its inter-annual variability and predictability. Seasonal climate forecasting (3-6 months ahead) has developed rapidly in recent years and their relevance to health decision makers is currently being explored [10]. There is now some evidence that large scale dust events in the Sahel are related to regional sea surface temperatures indicating that in the future they may be partially predictable (A. Ben Mohammed pers. comm.).

In order to clarify the mechanisms by which dust may contribute to meningitis epidemic occurrence and the means by which dust may be best monitored and predicted new research, data and models need to be developed. In particular a retrospective data set (minimum 20 years) of gridded estimates of dust (<2m height and a variety of particular sizes) would be helpful. The spatial scale of the data should be suitable for analysis at the scale of the average district in West Africa (with a focus on key countries such as Niger, Burkina and Mali) and the temporal scale should be aligned with that of the currently available health data (i.e. one week). It is important for the development of appropriate analysis that the level of uncertainty in the dust estimates is assessed and that the role of dust in relation to other environmental and climatic variables is understood. A careful analysis of the different sources of dust and their associated chemistry may provide valuable clues as to why certain areas are much more vulnerable to severe epidemics than others.

Here we propose that progress will only be made using a systematic approach, which expands the discussion between climate and health communities. The Meningitis Environmental Risk Information Technologies (MERIT) Consortium (http://merit.hc-foundation.org/) has evolved through the collaborative efforts of the World Health Organization (WHO) and members of the environmental, public health and research communities. The consortium aims to extend current capabilities to more effectively combine and use environmental information with knowledge of epidemic meningococcal meningitis through analysis that include of a range of relevant information and data on (for example) the spatial and temporal distribution of cases, population, environment and climate, vaccination status and strain characteristics

It is expected that this approach will have an immediate impact on public health outcomes in Africa through increasing the effectiveness of meningitis prevention and response control strategies.

References

[1] van Deuren, M., P. Brandtzaeg and J. W. M. van der Meer 2000 Update on meningococcal disease with emphasis on pathogenesis and clinical management. *Clinical Microbiology Reviews* **13**(1): 144-166.

[2] Stephens, D. S., B. Greenwood and P. Brandtzaeg 2007 Epidemic meningitis, meningococcaemia, and Neisseria meningitidis. *Lancet* **369**(9580): 2196-2210.

[3] Cuevas, L. E., I. Jeanne, A. Molesworth, M. Bell, E. C. Savory, S. J. Connor and M. C. Thomson 2007 Risk mapping and early warning systems for the control of meningitis in Africa. *Vaccine* **25**: A12-A17.

[4] Lapeyssonnie, L. 1963 La méningite cérébro-spinale en Afrique. Bulletin of the World Health Organisation 28(Supplement 1): 3-114.

[5] Molesworth, A. M., L. E. Cuevas, S. J. Connor, A. P. Morse and M. C. Thomson 2003 "Environmental risk and meningitis epidemics in Africa. *Emerging Infectious Diseases* **9**(10): 1287-1293.

[6] Thomson, M. C., A. M. Molesworth, M. H. Djingarey, K. R. Yameogo, F. Belanger and L. E. Cuevas 2006. Potential of environmental models to predict meningitis epidemics in Africa. *Tropical Medicine & International Health* **11**(6): 781-788.

[7] Molesworth, A. M., L. E. Cuevas, A. P. Morse, J. R. Herman and M. C. Thomson 2002 Dust clouds and spread of infection. *Lancet* **359**(9300): 81-82.

[8] Greenwood, B. M. 1987 The epidemiology of acute bacterial meningitis in tropical Africa. London, United Kingdom, Academic Press, Ltd.

[9] Roberts, L. (2008). An ill wind, bringing meningitis. *Science* **320**(5884): 1710-1715.

[10] Thomson, M. C., F. J. Doblas-Reyes, S. J. Mason, R. Hagedorn, S. J. Connor, T. Phindela, A. P. Morse and T. N. Palmer 2006. Malaria early warnings based on seasonal climate forecasts from multi-model ensembles. *Nature* **439**: 576-579.

African dust influence on ambient PM levels in South-Western Europe (Spain and Portugal): A quantitative approach to support implementation of Air Quality Directives

X Querol[1], A Alastuey[1], J Pey[1], N Pérez[1], M Escudero[1], S Castillo[1], Á Cristóbal[2], M Pallarés[2], A González[2], S Jiménez[2], N Alonso[2], S Alonso-Pérez[1,3], E Cuevas[3] and J de la Rosa[4]

[1] Institute of Earth Sciences "Jaume Almera", CSIC. Barcelona, Spain
[2] D.G. Calidad y Evaluación Ambiental, Ministerio de Medio Ambiente, Spain
[3] Izaña Atmospheric Research Center. Agencia Estatal de Meteorología. Tenerife, Spain.
[4] Universidad de Huelva, Spain

E-mail: xavier.querol@ija.csic.es

Abstract. This manuscript proposes and validates a methodology for the quantification of the daily African PM load during dust outbreaks in southern Europe. The daily net dust load in PM_{10} attributable to an African episode can be obtained by subtracting the daily regional background (RB) level from the PM_{10} concentration value at a RB station. The daily RB level can be obtained by applying a monthly moving 30th percentile to the PM_{10} time series at a RB station after a prior extraction of the data coincident with African dust transport. For days with influence of African dust, the dust load is given by the difference between the daily PM_{10} values minus the daily PM_{10} RB levels. This method allows us to quantify the net African dust load without chemical speciation. The comparison between the estimated net load during African dust outbreaks (ADO) and the crustal load determined by chemical speciation of PM_{10} filters at three RB stations in Spain had resulted in a very good correlation (R^2=0.60-0.83), being the equivalence (correlation lines' slopes ~ 1) highly significant in the three cases.

1. Introduction

The determination of the levels of atmospheric particulate matter (PM) is a key parameter in the evaluation of air quality given the proven influence of this pollutant on human health [1] and climate ([2] and references therein). As regards the adverse health effects of PM, the European Commission (EC) issued the Directive 1999/30/EU on air quality. This Directive established PM_{10} as a parameter to be monitored, with annual (40 μg m^{-3}) and daily (50 μg m^{-3}, not to be exceeded more than 35 occasions per year) limit values.

In cases where the exceedances of the limit values can be attributed to natural phenomena such as the transport of natural particles from arid regions, they can be discounted after scientific validation (article 2.15 from the 1999/30/EC Directive).

In southern Europe, the ADO may have a great impact on the PM levels [3,4,5,6,7]. Thus, at the Spanish EMEP (Co-operative Programme for Monitoring and Evaluation of the Long-Range Transmission of Air pollutants in Europe) sites more than 70% of the exceedances of the daily PM_{10}

© 2009 IOP Publishing Ltd

limit value are attributable to African dust contributions [8]. However, during ADO the PM_{10} fraction is not entirely composed of African crustal aerosols. [4] estimated that between 20-50% of the PM_{10} concentration at RB sites during ADO in NE Spain had a regional or local origin different from mineral dust. In addition, as discussed in [8], the formation of secondary aerosols is enhanced during ADO due to the interaction of local anthropogenic species with mineral dust components.

The most common approach to quantifying dust contribution is the development of chemical speciation studies. The main disadvantages of chemical speciation are the high economic cost and the length of the process.

In the light of the foregoing discussion, we propose a quantitative and simple methodology for estimating the net African dust load in PM_{10} during ADO. This methodology has been validated by chemical speciation data for three RB sites in Spain.

2. Methods

2.1. Measurements and interpretation

The variability of PM_{10} levels from thirteen RB stations was interpreted. The air quality monitoring stations selected are representative of the different climatic areas of the Iberian Peninsula (table 1 and figure 1). PM_{10} measurements at EMEP stations were performed by means of the gravimetric method making use of MCV-PM1025 samplers (30 m^3 h^{-1}). At Valderejo and Izki, PM_{10} measurements have been carried out using β-attenuation instrumentation and, at Monagrega, PM_{10} levels have been obtained with TEOM instrumentation. PM_{10} real time levels were corrected through the application of the correction factors obtained by comparison with the gravimetric measurements.

Table 1. Spanish RB stations providing PM_{10} data.

Station	Air quality monitoring network	Location	Altitude (m.a.s.l.)	PM_{10} monitoring	Sampling period (PM_{10} Chemical Speciation)
O Saviñao	EMEP	42° 38' N 07° 42' W	506	Gravimetric	2001-2003
Niembro-Llanes	EMEP	43° 26' N 04° 51' W	134	Gravimetric	2001-2003
Valderejo	Government of the Basque Country	42° 53' N 03° 14' W	911	β-Attenuation	Since 2000
Izki	Government of the Basque Country	42° 39' N 02° 30' W	835	β-Attenuation	Since 2001
Cabo de Creus	EMEP	42° 19' N 03° 19' E	23	Gravimetric	2001-2003
Els Torms	EMEP	41° 24' N 00° 43' E	470	Gravimetric	2001-2003
Monagrega	ENDESA	40° 59' N 00° 12' W	600	TEOM	Since 1996 (24/03/1999-06/07/2000)
Risco Llano	EMEP	39° 31' N 04° 21' W	1241	Gravimetric	2001-2003
Campisábalos	EMEP	41° 17' N 03° 09' W	1360	Gravimetric	2001-2003
Peñausende	EMEP	41° 17' N 05° 52' W	985	Gravimetric	2001-2003
Zarra	EMEP	39° 05' N 01° 06' W	885	Gravimetric	2001-2003
Barcarrota	EMEP	38° 29' N 06° 55' W	393	Gravimetric	2001-2003
Víznar	EMEP	37° 14' N 03° 28' W	1265	Gravimetric	2001-2003
Montseny	CSIC-Generalitat de Catalunya Goverment	41° 46' N 02° 21' E	720	Laser Spectrometer	Since 2002 (Since 2002)
Castillo de Bellver	Illes Balears Goverment	39° 34' N 02° 37' E	117	β-Attenuation	2004-2005 (08/01/2004-29/97/2005)

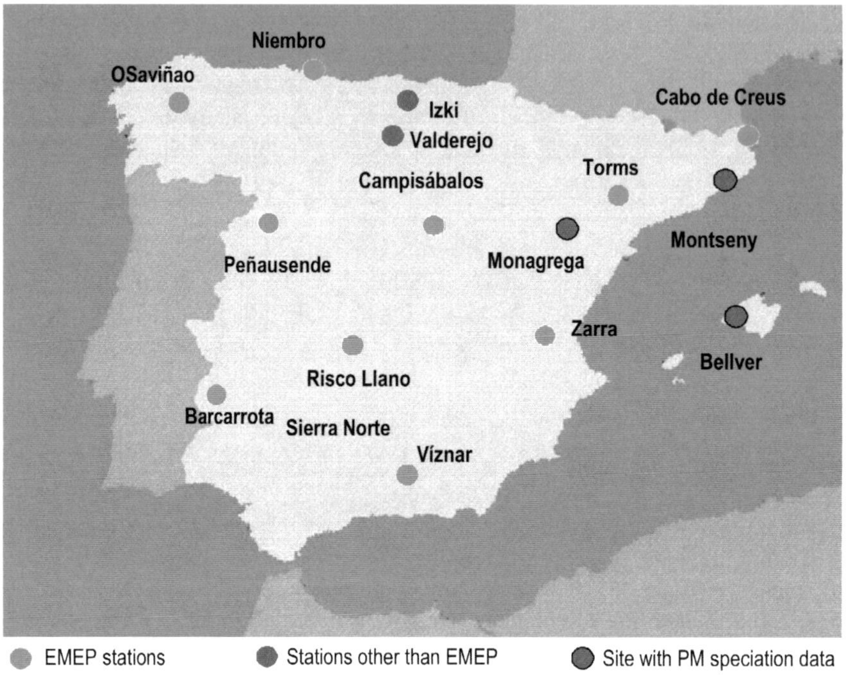

Figure 1. Spanish RB stations providing PM_{10} data and/or chemical speciation data.

The interpretation of the origin of the air masses reaching each of these monitoring sites was performed by calculating back-trajectories with the HYSPLIT4 model [9] accomplished with other tools such as meteorological maps, numerical models, and satellite images (see details in [5]).

2.2. PM_{10} speciation

In order to validate the methodology proposed here, data on chemical speciation of PM_{10} from three RB sites located in the Iberian Peninsula and the Balearic Islands were used (Monagrega, Montseny and Castillo de Bellver, table 1 and figure 1). 24-hour samples of PM_{10} were collected in quartz fibre filters twice per week. The instruments for the gravimetric samplings were high volume samplers working at 30 m^3 h^{-1}. At Monagrega and at Montseny, DIGITEL inlets were used whereas MCV-PM1025 instrumentation was used at Castillo de Bellver.
Once PM_{10} levels were gravimetrically determined, the filters were analyzed by different techniques to quantify the concentrations of approximately 60 PM components [10]. The concentrations of the major crustal components (SiO_2, Al_2O_3, CO_3^{2-}, Ca, Mg, Fe and K) accounted for >99% of the mineral dust load.

3. Results

3.1. Identification of African episodes affecting PM levels

The identification of the origin of the air masses reaching the regional monitoring sites and the interpretation of the origin of PM events was carried out following the above methodology (see details in [5]).

The persistence of some African dust episodes has been taken into account. It should be noted that PM_{10} levels may be high at RB stations one day before and/or two days after the period with proven influence of African dust. This could arise when the African air masses transported towards the Iberian Peninsula are not followed by episodes of intense advection. According to other studies, African dust transported towards southern Europe reaches a minimum altitude of 1500 m.a.s.l. [11]. The deposition of particles between 1 and 10 microns has an average speed of 0.6 cm s^{-1} [12]. Thus, levels of PM could remain high at a given site two days after the episode had ended and, therefore, the African period could also include these dates.

3.2. Calculation of the RB levels

During Atlantic advection episodes in Spain, PM_{10} levels are at their lowest owing to the reduced external contributions. At an specific RB site, the mean moving average of Atlantic advection can be considered as similar to the RB PM_{10} levels. This was performed at the 13 RB sites (table 1 and figure 1) for the period 2001-2003. The execution of this procedure requires daily interpretation of the air masses origin, and therefore, is not a straightforward method to be applied by the air monitoring networks.

In order to simplify the calculation of the RB PM_{10} levels, we propose the application of a monthly moving percentile to the PM_{10} series, after excluding the data corresponding to the days with ADO. This method is easier to apply given that only the days with influence of African dust transport need to be identified. In order to select the most adequate percentile, we performed a sensitivity test using different percentile thresholds (5, 10, 30, 40 and 50th) at the 13 RB stations. As shown in figure 2, the application of the 10 and 30th percentiles (y=1.44x and y=1.07x, respectively) underestimates the RB levels when compared with those calculated after considering only the days with Atlantic advection. By contrast, the use of the 40 and 50th percentiles yielded an overestimation of the RB levels (y=0.96x and y=0.86x, respectively). Thus, the 30th and 40th percentiles proved to be the most suitable. In order to reduce the underestimation of the African dust load we propose the adoption of the 30th percentile.

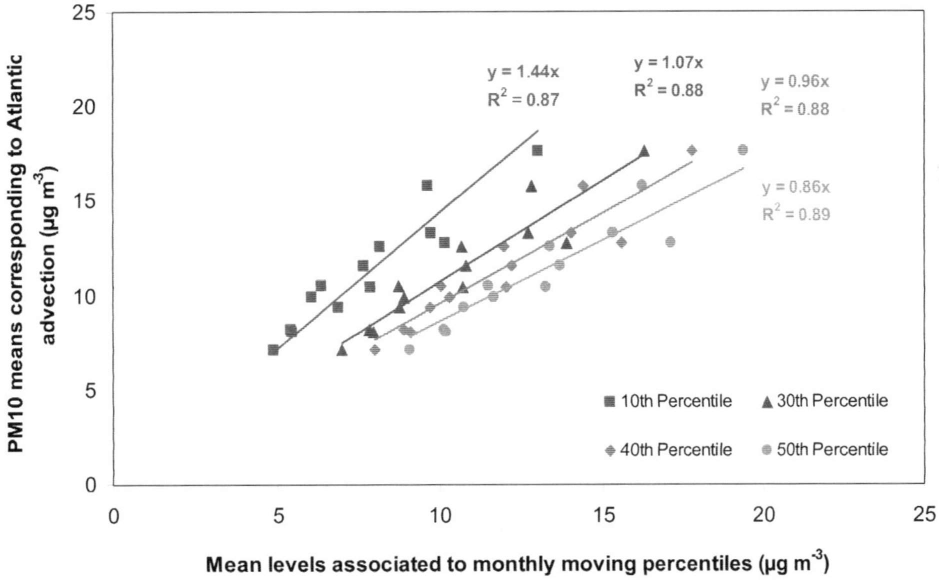

Figure 2. Correlation between the PM_{10} averages calculated only for the days with Atlantic advection (ATL) and the means of the time series obtained after applying different monthly moving percentiles (5th, 10th, 30th, 40th and 50th) to the PM_{10} data from 13 RB stations of the Iberian Peninsula for the period 2001-2003.

3.3. Calculation of the net African dust load and validation

The daily net African dust load was calculated by subtracting the daily RB level (obtained with the 30th percentile) from the daily PM_{10} value registered during an ADO.

In order to validate this method, we compared the mineral matter (sum of SiO_2, Al_2O_3, CO_3^{2-}, Ca, Mg, Fe and K) load for days with ADO from three RB monitoring stations (Monagrega, Montseny and Castillo de Bellver, Figure 1) with the net dust contribution calculated by using the proposed method. As shown in figure 3, the crustal load determined by chemical speciation strongly correlated ($0.60<R^2<0.69$) with the net African dust load calculated. Therefore, the net African dust load calculated is mainly composed of mineral matter, which corroborates the major natural origin of the PM_{10} fraction, although sulphation and nitration of dust is evidenced in specific cases.

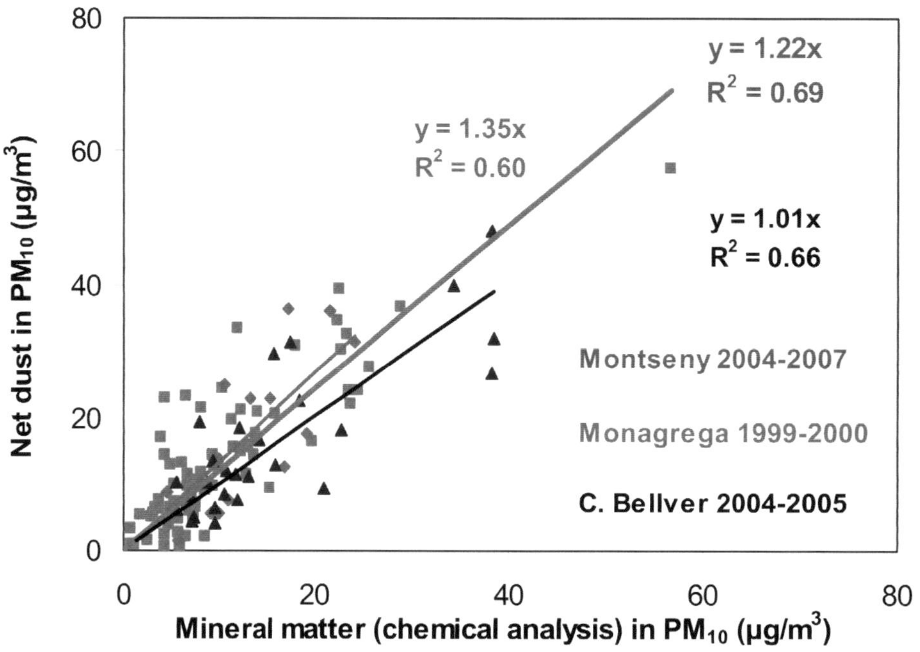

Figure 3. Correlation between the net African dust load obtained with the proposed methodology and the crustal load in PM_{10} determined by chemical speciation at Monagrega, Montseny and Castillo de Bellver.

3.4. Application of this method during exceedences of the daily limit value of PM_{10}

When the PM_{10} daily limit value is exceeded at urban or industrial stations during an ADO, the net African dust load can be used to evaluate the impact of the natural contribution to that exceedance. For example, at two stations, urban background and kerbside, 60 and 100 $\mu gPM_{10}\ m^{-3}$ are measured, respectively, on a given day during an ADO. Simultaneously, at the closest RB site, a daily value of 41 $\mu gPM_{10}\ m^{-3}$ is measured, and the monthly moving 30th percentile value is 10 $\mu g\ m^{-3}$ for this day. Therefore, the net African dust load in this region, for this day, is 31 $\mu gPM_{10}\ m^{-3}$. In this case, the exceedance at the urban background site can be attributed to the dust contribution (60-31 = 29 $\mu gPM_{10}\ m^{-3}$ as anthropogenic load), whereas the at the kerbside station must be ascribed to anthropogenic contributions (100-31 = 69 $\mu gPM_{10}\ m^{-3}$).

In some cases, negative values could be obtained when subtracting the net African dust load. In these situations, the net African dust load should be calculated at another RB station in the vicinity of the urban or industrial station at approximately the same altitude.

Acknowledgements
This study was supported by the S.G.D. Environmental Quality and Industrial Environment from the Spanish Ministry of the Environment, but also by research projects from the Ministry of Science and Innovation (CGL2005-03428- C04-03/CLI, CGL2007-62505/CLI, GRACCIE- CSD2007-00067), the European Union (6th framework CIRCE IP, 036961, EUSAAR RII3-CT-2006-026140) and the Ministry of the Environment (CALIOPE, 441/2006/3-12.1). The authors would like to express their gratitude to the Atmospheric Modelling & Weather Forecasting Group in the University of Athens, the Earth Sciences Department of the Barcelona Supercomputing Center, the NASA/Goddard Space Flight Center, NOAA Air Resources Laboratory (ARL), the Naval Research Laboratory and the SeaWiFS project (NASA) for the provision of the SKIRON, DREAM, TOMS, NAAPs aerosol maps and the satellite imagery, respectively. The authors are also indebted to the Spanish Ministry of the Environment for providing data from EMEP air quality network, the Autonomous Government of the Basque Country for supplying data from Valderejo and Izki stations and ENDESA-Teruel for providing data from Monagrega station. Finally, the authors gratefully acknowledge the NOAA Air Resources Laboratory (ARL) for the provision of HYSPLIT model.

References
[1] Pope C.A. and Dockery D.W. 2006 Health effects of fine particulate air pollution: lines that connect, A *Journal of the Air & Waste Man. Ass.,* **56**, 709-742.
[2] IPCC, 2007. Climate Change 2007: The Physical Science Basis. Contribution of Working Group I to the Fourth Assessment Report of the IPCC (ISBN 978 0521 88009-1 Hardback; 978 0521 70596-7 Paperback).
[3] Querol X., Alastuey A., Puicercus J.A., Mantilla E., Miró J.V., López-Soler A., Plana F. and Artíñano B. 1998 Seasonal Evolution of Suspended Particles Around a Large Coal-Fired Power Station: Particle Levels and Sources, *Atm. Env.,* **32**, 1963-1978.
[4] Rodríguez S., Querol X., Alastuey A. and Plana F. 2002 Sources and processes affecting levels and composition of atmospheric aerosol in the Western Mediterranean, *Journal of Geophysical Research,* **107** (D24), 4777.
[5] Escudero M., Castillo S., Querol X., Avila A., Alarcón M., Viana M.M., Alastuey A., Cuevas E. and Rodríguez S. 2005 Wet and dry African dust episodes over Eastern Spain, *Journal of geophysical research,* **110** (D18S08), 10.1029.
[6] Gerasopoulos E., Kouvarakis G., Babasakalis P., Vrekoussis M., Putaud J.P. and Mihalopoulos N. (2006). Origin and variability of particulate matter (PM10) mass concentrations over the Eastern Mediterranean, Atm. Env., **40**, 4679-4690.
[7] Gobbi G.P., Barnaba F. and Ammannato L. 2007 Estimating the impact of Saharan dust on the year 2001 PM10 record of Rome, Italy, *Atm. Env.,* **41**, 261-275.
[8] Escudero M., Querol X., Avila A. and Cuevas E. 2007 Origin of the exceedances of the European daily PM limit value in regional background areas of Spain, *Atm. Env.,* **41**, 730-744.
[9] Alastuey A., Querol X., Castillo S., Escudero M., Avila A., Cuevas E., Torres C., Romero P.M., Expósito F., Garcia O., Diaz J.P., Dingenen R.V. and Putaud J.P. 2005 Characterisation of TSP and $PM_{2.5}$at Izaña and Sta. Cruz de Tenerife (Canary Islands, Spain) during a Saharan dust episode (July 2002), *Atm. Env.*, **39**, 4715-4728.
[10] Draxler R.R. and G.D. Rolph 2003 HYSPLIT (HYbrid Single-Particle Lagrangian Integrated Trajectory). Website: http://www.arl.noaa.gov/ready/hysplit4.html, NOAA Air Resources Laboratory, Silver Spring, MD.
[11] Querol X., Alastuey A., Rodríguez S., Plana F., Ruiz C.R., Cots N., Massagué G. and Puig O. 2001 PM_{10} and $PM_{2.5}$ source apportionment in the Barcelona Metropolitan Area, Catalonia, Spain, *Atm. Env.,* **35**, 6407-6419.

[12] Pérez C., Sicard M., Jorba O., Comeron A. and Baldasano J.M. 2004 Summertime re-circulations of air pollutants over the north-eastern Iberian coast observed from systematic EARLINET lidar measurements in Barcelona, *Atm. Env.,* **38**, 3983-4000.
[13] Zender C.S., Bian H.S. and Newman D. 2003 Mineral Dust Entrainment and Deposition (DEAD) model: Description and 1990s dust climatology, *Journal of geophysical research,* **108** (D14), Art. No. 4416.

Short-Term Changes in the Northwest African Upwelling System induced by Saharan Dust Deposition events

A G Ramos[1], E Cuevas[2], C Pérez[3], J M Baldasano[3,4], J Coca[1], A Redondo[1], S Alonso-Pérez[2,5], J J Bustos[2] and S Nickovic[6]

[1]SeaSnet Canarias. Dpto. de Biología (Univ. of Las Palmas de Gran Canaria). Canary Islands, Spain.
[2]Izaña Atmospheric Research Center. Agencia Estatal de Meteorología. Tenerife, Spain.
[3]Earth Sciences Department. Barcelona Supercomputing Center. Barcelona, Spain.
[4]Laboratory of Environmental Modeling. Universitat Politècnica de Catalunya. Barcelona. Spain
[5] Institute of Earth Sciences "Jaume Almera", CSIC. Barcelona, Spain.
[6]Atmospheric Research and Environment Branch. World Meteorological Organization. Geneva, Switzerland.

E-mail: aramos@pesca.gi.ulpgc.es

Abstract. During the last 7-year period (2000-2006) atmosphere circulation changes show strong influences on the dust storm deposition dynamics and, as a result, on the primary production dynamics of the northwest African Upwelling System. From 2000 to 2006, the annual mean sea level pressure became higher ranging from 1014 to 1015 mb. Mean annual zonal wind intensity became higher (from 1.1 to 1.8 m s^{-1}), while the mean annual meridional was reduced from 6.2 to 5.3 m s^{-1} at the north of the Canary Islands. Mean annual satellite-derived AVHRR/NOAA SST recorded in the northwest African Upwelling became warmer in both locations, from 18.3°C to 18.8°C in Cape Ghir and from 19.5°C to 20.3°C north Canary Islands waters. CHL records from the SeaWiFS/OV-2 showed a different pattern trend. Mean annual CHL levels increased at Cape Ghir from 0.65 mg m-3 to 0.9 mg m-3 and significantly reduced from 0.59 mg m^{-3} to 0.31 mg m^{-3} at the north of the Canary Islands. Changes observed in the role of CHL during the last 7-years period could be associated to intensive dust deposition and exceptional weather warming observed in this area since 2000. However, this study focused on a 7-year period and conclusions on possible links between dust deposition and marine biochemistry activity cannot be generalized.

1. Introduction

Photosynthesis in the ocean is, by far, the most important biochemical process of conversion of Inorganic Carbon (IC) into biomass (Organic Carbon, OC). The OC generated through this primary production (photosynthetic) process is transferred to the higher trophic layer in superficial waters where most marine life develops. However, the excess of unused C (rejected) undergoes a progressive sinking to the bottom. In the deep ocean layers, this rejected OC can go through two processes: it is either recycled by the heterotrophic flora and transformed into reusable IC or it stays fixed as a sedimentary stratum at the bottom [1]. This cycle, however, requires some additional nutrients in high

© 2009 IOP Publishing Ltd

concentrations, mainly Nitrogen (N) and Phosphorus (P). The availability (upwelling system) or non-availability (oligotrophic oceanic waters) of any of these macronutrient elements (N and P) depends on the vertical mixing supply to upper waters or by diazotrophic nitrogen-fixing processes.

However, the absence of other available key nutrients such as Iron (Fe) and Manganese (Mn) can also limit the conversion of IC into OC. This micronutrient limitation is associated to its extremely low solubility (nanomolar levels) and are mostly supplied in the NE Atlantic by the exogenic dust deposition events [2,3,4,5,6].

Thus, long-term trend changes on the concentration of these trace elements would change the primary production role in both poor oligotrophic and rich eutrophic oceanic areas. In this paper, we focus in preliminar long-term trend of geophysical and meteorological indicators in the northwest African waters during the warmest 7-years period of the century (2000-2006), and how this changes influence the dust deposition input and, for instance, the primary production role in both areas.

2. Methodology

Meteorological parameters, satellite images data and model dust deposition inputs were collected from january 2000 to December 2006 over the northwest African waters at two stations located at Cape Ghir (upwelling station) and north Canary Island (figure 1).

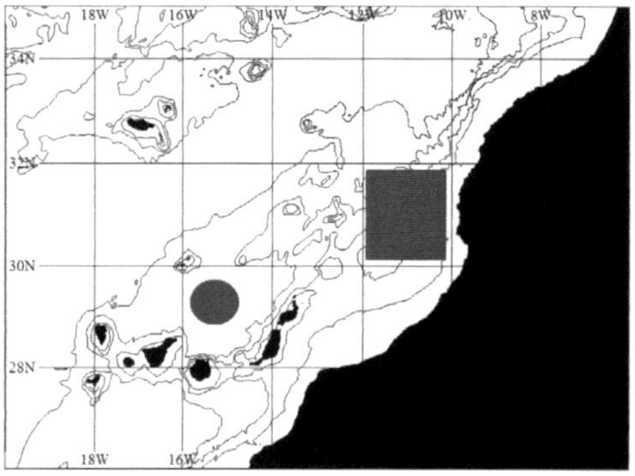

Figure 1. The northwest African Upwelling area. Data sets of dust deposition, SST and CHL were extracted from oligotrophic waters located to the north of the Canary Islands (grey circle) from upwelled waters located off Cape Ghir (grey square).

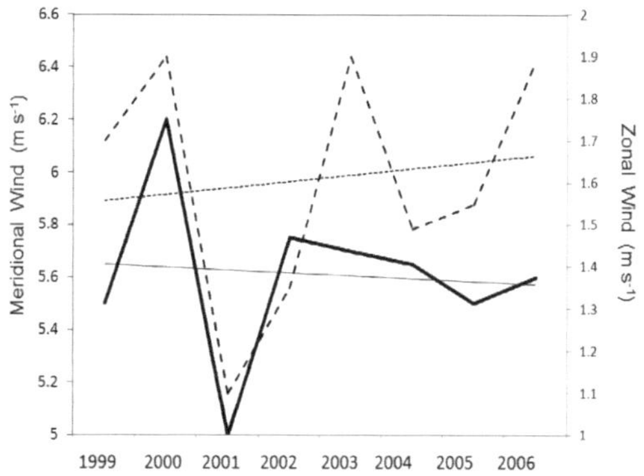

Figure 2. Annual mean meridional (solid line) and zonal (dashed line) wind speed series in the Northwest African Upwelling area from 1999 to 2006.

Wind components at 10 m height were obtained from the European Centre for Medium-Range Weather Forecasts (ECMWF) T511 L60 operational analysis at 00, 06, 12 and 18 UTC. We used the Dust REgional Atmospheric model (DREAM) of the Barcelona Supercomputing Center (BSC) [7,8,9,10] to simulate predict the atmospheric cycle of mineral dust aerosol. The model incorporates advanced parameterizations of all the major phases of the atmospheric dust cycle, such as emission, diffusion, advection and deposition. All the model components are described in [11]. Satellite-derived parameters were obtained during the event: Sea Surface Temperature (SST) images were supplied by the Jet Propulsion Laboratory (JPL) PODAAC Pathfinder. Satellite-derived SeaWiFS/OV-2 chlorophyll-a (CHL) images were extracted from the NASA Ocean Colour Web. Data were projected on SMI cylindrical equidistant at different ground resolutions [12].

3. Results and discussion

Mean sea level pressure, AVHRR/NOAA SST and wind components obtained from the ECMWF showed the warmest 7-year period reported since 1912. From 2000 to 2006, the annual mean sea level pressure was higher ranging from 1014 to 1015 mb. This change on the atmospheric circulation influenced the zonal and meridional wind components dynamics. Mean annual zonal wind intensity became higher during the 7-year period (from 1.1 to 1.8 m s^{-1}). However, the mean annual meridional winds were reduced from 6.2 to 5.3 m s^{-1} (figure 2).

The increase of the east-west component of the wind together with a reduction of the north trade wind component was translated into a change of the annual flux of Saharan dust over the northwest African Upwelling system. Annual average dry deposition flux of Saharan dust can range from 0.03 to 0.08 g m^{-2} d^{-1} or $1.7x10^6$ t yr^{-1} [13]. Our DREAM model derived data showed that at the north Canary Islands station, the annual dry dust deposition data reduced from 0.12 mg m^{-2} d^{-1} (2000) to 0.10 g m^{-2} d^{-1} (2006) north Canary Islands. However, in the Cape Ghir location station DREAM model derived data showed that annual dry dust deposition data increased from 0.025 mg m^{-2} d^{-1} (2000) to 0.040 mg m^{-2} d^{-1} (2006) (figure 3).

Mean annual satellite-derived AVHRR/NOAA SST recorded in the northwest African Upwelling becomes warmer. It ranged between 18.3°C (2000) to 18.8°C (2006) in the coastal eutrophic station located at Cape Ghir and from 19.5°C (2000) to 20.3°C (2006) in the oceanic oligotrophic station north Canary Islands waters (figure 4). Some authors [Ramos et al., 2005] revealed that in the Canarian waters, the SST ranges between 17° C in winter to 24° C in summer. However, some areas at the south of the Canarian waters presented SST peaks during this period (2004) of 28.5° C, records never reported before.

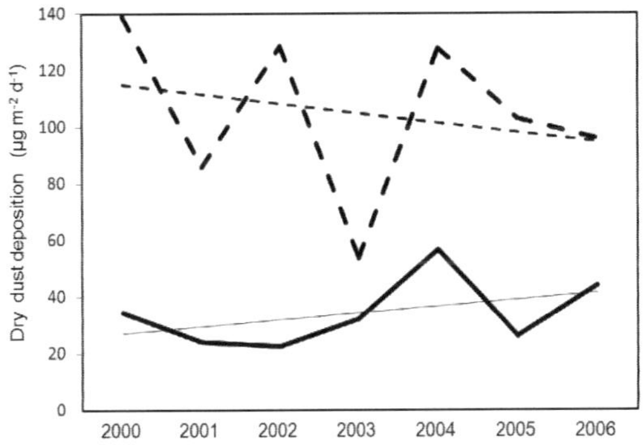

Figure 3. Annual mean DREAM modelled dry dust deposition at Cape Ghir (solid line) and north of the Canary Islands (dashed line) from 2000 to 2006.

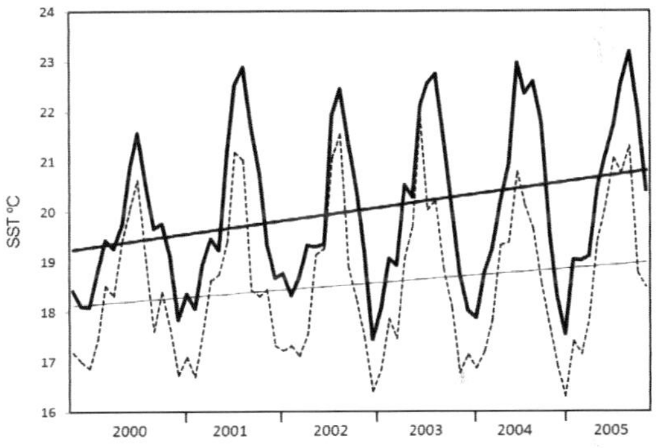

Figure 4. Annual mean satellite-derived AVHRR/NOAA SST data series at Cape Ghir (dashed line) and north of the Canary Islands (solid line) from 2000 to 2005.

Some studies [1,14] investigated the mean annual cycles of CHL in this area, concluding that in the canarian oligrotrohic waters, values of mean annual CHL oscillated around 0.4 mg m^{-3}. However, in the rich eutrophic upwelled waters of Cape Ghir the mean annual CHL levels arised to 0.8 mg m^{-3}. As a result of the progressive increase of the SST observed in both Canary Island and Cape Ghir waters during 2000-2006, it could be expected a progressive reduction of the mean annual CHL concentration in both locations. Our CHL images from the SeaWiFS/OV-2 showed a different pattern trend. Mean annual CHL levels increased (like dry dust deposition) at Cape Ghir from 0.65 mg m^{-3} (2000) to 0.9 mg m^{-3} (2006) indicating a process of progressive eutrophization on the upwelling sector. However, in the Canary Islands waters the mean annual CHL levels becomes significantly reduced from 0.59 mg m^{-3} (2000) to 0.31 mg m^{-3} (2006) (like dry dust deposition), indicating a process of progressive tropical-like ocean landscape (figure 5).

Annual fluxes changes of Saharan dust could produce a significant impact on the bigeochemical cycle of trace elements, providing a source of nutrients required for N2-fixing diazotrophic cyanobacterial growth [3,5]. Thus, [15] reported in 2004 the development of an extensive and inedit bloom of the diazotrophic cyanobacterium Trichodesmium erythraeum Ehrenberg in the northwest African Upwelling never described before [16]. The early stages of this anomalous diazotrophic event and its possible relationship to episodic Saharan dust deposition may represent the dominant source of these new CHL pool observed in the Upwelling System location. However, in oceanic oligotrophic waters the reduction of the dust fluxes observed in this work could generate a significant reduction of the CHL levels and as a direct consequence, this sector would become poorer.

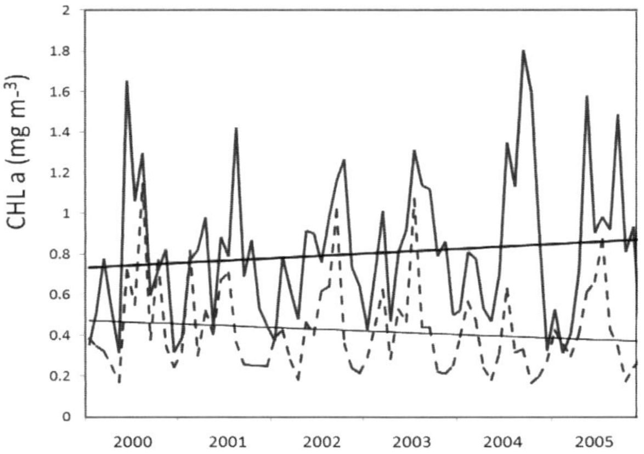

Figure 5. Annual mean satellite-derived SeaWiFS/OV-2 CHL data series obtained at Cape Ghir (solid line) and north of the Canary Islands (dashed line) in the Northwest African Upwelling area from 2000 to 2005.

4. Conclusions

We hypostatize that the apparent changes observed in CHL during the last 7-year period could be associated to intensive dust deposition and exceptional weather warming observed in this area since 2000. This study is addressed to a 7-year period and our conclusions on possible links between dust deposition and marine biochemistry activity cannot be generalized. It would be necessary the study of iron bioavailability by using results of an atmospheric iron model (currently being under development) which simulates the atmospheric processing of iron and its conversion from insoluble to soluble form for primary producers.

References

[1] Basterretxea, G. and Arístegui, J, 2000 Mesoscale variability in phytoplankton biomass distribution and photosynthetic parameters in the Canary - NW African coast transition zone. *Marine Ecology Progress Series,* **197**, 27-40.

[2] Prospero, J. M. and Carlson, T. N. 1981 Saharan air outbreaks over the tropical North Atlantic. *Pure and Applied Geophysic,s* **119** (3), 677-691.

[3] Lenes, J., Darrow, B., Cattrall, C., Heil, C. A., Callahan, M., Vargo, G., Byrne. R., Prospero, J., Bates, D., Fanning, K. and Walsh, J. 2001 Iron fertilization and the Trichodesmium response on the West Florida shelf. *Limnology and Oceanography,* **46**, 1261-1277.

[4] Liu, X.and Millero, F.J. 2005 The solubility of iron in seawater. *Marine Chemistry* ,**77**, 43-54.

[5] Capone, D.G, Zehr, J.P, Paerl, H.W, Bergman, B., and Carpenter, E.J 1997 Trichodesmium, a globally significant marine cyanobacterium. *Science,* **276**,1221-1229.

[6] Kremling, K. and Streu, P., 1993 Saharan dust influenced trace elements fluxes in deep North Atlantic subtropical waters. *Deep Sea Research,* Part I, **40**, 1155-1168.

[7] Pérez, C., Nickovic, S., Baldasano, J.M., Sicard, M., Rocadenbosch, F. and Cachorro, V.E. 2006 A long Saharan dust event over the western Mediterranean: Lidar, Sun photometer observations, and regional dust modelling. *Journal of Geophysical Research,* **111**, D15214, doi:10.1029/2005JD006579.

[8] Pérez, C., Nickovic, S., Pejanovic, G., Baldasano, J.M., and Özsoy, E., 2006 Interactive dust radiation modelling: A step to improve weather forecasts. *Journal of Geophysical Research* , **111**, D16206, doi:10.1029/2005JD006717.

[9] Papayannis, A., Zhang, H.Q., Amiridis, V., Ju, H.B., Chourdakis, G., Georgoussis, G., Pérez, C., Chen, H.B., Goloub, P., Mamouri, R.E., Kazadzis, S., Paronis, D., Tsaknakis, G., and Baldasano, J.M., 2007 Extraordinary dust event over Beijing, China, during April 2006: Lidar, Sun photometric, satellite observations and model validation. *Geophysical Research Letters,* **34**, L07806, doi:10.1029/2006GL029125.

[10] Pérez, C., Jiménez-Guerrero, P., Jorba, O., Baldasano, J.M., Cuevas, E., Nickovic, S., Querol, X., 2007 Long-term simulations (1958-2006) of Saharan dust over the Mediterranean and the Eastern North Atlantic with the DREAM regional dust model. In: XXIV International Union of Geodesy and Geophysics (IUGG) General Assembly, Perugia, Italy, 2-13 July 2007.

[11] Nickovic, S., Kallos, G., Papadopoulos, A., and Kakaliagou, O., 2001 A model for prediction of desert dust cycle in the atmosphere. *Journal of Geophysical Research,* **106**, D16, 18113–18129.

[12] Coca, J. and Ramos, A., 2004 Relationships between satellite-derived oceanic events and the albacore tuna (Thunnus alalunga, Bonaterre, 1788) artisanal fishing grounds in the NE Atlantic. In Remote Sensing of the Ocean and Sea Ice (Ed. C. Bostater Jr. and Rosalia Santoreli). Proceedings of the SPIE, 5569, 116-129.

[13] Torres, M., Gelado, M., Collado, C., Siruela, V., Cardona, P., and Hernández-Brito, J., 2002 Variability of dust inputs to the CANIGO zone. *Deep-Sea Research II*, **49**, 3455-3464.

[14] Arístegui, J., Tett, P., Hernández-Guerra, A., Basterretxea, G., Montero, M.F., Wild, K., Sangrá, P., Hernández-León, S., Cantón, M., García - Braun, J.A, Pacheco, M., and Barton, E.D., 1997 The influence of island-generated eddies on chlorophyll distribution: a study of mesoscale variation around Gran Canaria. *Deep-Sea Research*, **44**, 71-96.

[15] Ramos A., Martel, A., Codd, G., Soler, E., Coca, J., Redondo, A., Morrison, L., Metcalf, J., Ojeda, A., Suárez, S., and Petit. M., 2005 Bloom of the marine diazotrophic cyanobacterium Trichodesmium erythraeum Ehrenberg in the Northwest African Upwelling. *Marine Ecology Progress Series*, **301**, 303-305.

[16] Hood, R.R., Subramaniam, A., May, L.R., Carpenter, E.J. and Capone, D.G. 2002 Remote sensing of nitrogen fixation by Trichodesmium. *Deep-Sea Research II,* **49**, 23-147.